Teaching design and technology in the primary school

Tina Jarvis

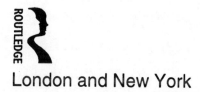

London and New York

First published 1993
by Routledge
11 New Fetter Lane, London EC4P 4EE

Simultaneously published in the USA and Canada
by Routledge
29 West 35th Street, New York, NY 10001

Typeset in Garamond by
NWL Editorial Services, Langport, Somerset

Printed and bound in Great Britain by
T.J. Press (Padstow) Ltd, Padstow, Cornwall

British Library Cataloguing in Publication Data
A catalogue reference for this book is available from
the British Library

ISBN 0–415–07228–X

Library of Congress Cataloging in Publication Data
Jarvis, Tina.
   Teaching design and technology in the primary school /
   Tina Jarvis.
      p.   cm.
   ISBN 0–41507228–X
   1. Design, Industrial – Study and teaching (Elementary) 2.
Technology – Study and teaching (Elementary)   I. Title.
   TS171.4J37   1993   92–45849
   372.3'58 – dc20   CIP

Teaching design and technology in the primary school

KING ALFRED'S

Design and technology is a new subject. Many teachers have little idea what it is or what they are trying to achieve with it. While a great deal is available on how to make things, *Teaching Design and Technology in the Primary School* alone at the moment tells them how to put together a meaningful curriculum in the subject. Old CDT courses were purely secondary whereas D & T is an academic National Curriculum subject taught from age 5 and this book provides guidance on strategies for including it effectively within the whole curriculum. It covers the development of cooperative group-work, finding effective ways to assess individuals in group situations, and looks at how teachers can tackle subject areas which may be unfamiliar to them, such as energy and machines, systems and environments.

Tina Jarvis taught in several inner-ring Birmingham primary schools before becoming Lecturer in Primary Science and Technology at the University of Leicester. Her publications include *Children and Primary Science* and several articles on science and technology.

Many thanks to Frieda Billingham for all her assistance and invaluable comments. I would also like to thank my husband Peter for all his help and support.

Contents

Figures and tables

FIGURES

TABLES

Chapter 1

Why teach design and technology?

'Technological product' covers virtually any man-made item, since technology is the application of knowledge and skills to create processes and products that meet human needs. These products are not limited to artifacts but also include systems and environments and can be made of many materials, including textiles, constructional materials and food.

A system can be the arrangement of two or more artifacts or organisations of people to perform a task. A bicycle can be described as the combination of wheels, gears, braking mechanism, etc. to enable someone to travel more efficiently, and an orchestra is another system where both people and artifacts combine to produce a piece of music. A man-made environment may include the interior design of a home or school, a garden, layout of a farm, shopping complex or even the arrangement of a fish tank.

As recognised during the 1993 Technology National Curriculum consultation process, links with science are particularly important in creating structures and mechanical devices as well as controlling systems. However, in order to appreciate and understand technology fully, design and making activities need to be practiced in many curricular areas and contexts. The wide scope and possibilities of technology can be imaginative and exciting. However, they must be rather concerning to the primary teacher who is already working extremely hard to implement the requirements of the Core subjects in addition to the other foundation subjects.

It is important, therefore, that teachers see that design and technology has merit and potential not offered by other subjects and that it can be effectively integrated into the existing primary curriculum to the advantage of children's learning and development. In the long term design and technology has the potential to help create imaginative, thinking, tolerant and responsible adults better able to adapt and cope with day-to-day problems and aware of the effect of their actions. In the short term it can enhance other subjects in the primary curriculum by showing their relevance in actual situations and stimulating children's enthusiasm through its very practical nature.

At its basic level design and technology is about improving day-to-day survival in order to find effective ways of providing food, water, warmth, shelter, security, clothing, health and opportunities for interacting with others. As societies develop so these refinements are more related to enriching the quality of life. The process in each case is similar and is outlined in the Design and Technology National Curriculum. This includes observation and evaluation of existing situations to identify needs and opportunities for improvement. The way to achieve these are explored and planned, an approach is decided on and implemented. The success of action is evaluated in terms of costs which include financial, time involved and the effect on others. The problem to be solved may be fairly minor such as reducing friction on a toy car or more formidable as in creating a more efficient national transport network. In each case action will often create further needs or difficulties to be overcome and so the cycle continues.

A foundation for adulthood

Design and technology, therefore, provides a valuable foundation for effective problem solving in adult life as everyone is frequently involved in technological activities in their day-to-day lives, although they might not describe the process as technology. Finding ways to organise out-of-school care for children; deciding on whether and how to improve a home; coping with the need to have suitable clothes for a special event; and making appropriate meals for particular individuals could all be described as technological tasks. The process advocated by the National Curriculum enables children to learn to tackle such tasks in an effective manner: considering a range of solutions, planning effectively, applying knowledge and skills to use a wide variety of materials and tools, and constantly reviewing and improving their approach.

In addition the design and technology curriculum also requires children to learn to evaluate manufactured products which should help them as adults to be aware of the significant criteria to be considered when buying or using new products and should discourage automatic acceptance of advertisements and other propaganda without critical consideration. Technological skills should also help children to become more informed about the implications and potential in developing or improving different enterprises and businesses.

In order to demonstrate the relevance of technological study to adult work, teachers could share instances when teaching can be seen as a technological activity with the children. Preparing a term's work requires that teachers identify what the class needs, teach the sessions and finally evaluate its success in terms of the children's learning in order to extend it in the following term. On a smaller scale the production of a classroom display is an artifact to meet the need of providing a stimulating classroom and to encourage the children, which involves planning within restrictions of finance, time, space, materials and children's abilities. Indeed almost every moment of the teacher's time involves evaluating and assessing each child's performance to identify their individual needs and then trying to find ways of responding to those needs in the most effective way possible, bearing in mind all the available facilities, support and limitations of staffing, finances and equipment.

Responding to rapid technological change

By developing the skills inherent in design and technology children should not only be better able to solve a wide variety of problems when they are adults, they should also become more aware of the way technology is affecting society by changing the home, work place, and lifestyles and be able to evaluate it. They should learn that technological change cannot easily be reversed. Ideally they will understand its great power and start to appreciate their responsibilities in its process. If children are able to understand the process of technological change they will not just unquestioningly accept it but feel that they can challenge and alter it.

Developing independence and responsibility

In order to develop the ability to identify needs and take appropriate action children should be involved in some of the decisions about what they do in the school. They might help to decide how to set up a central storage area for tools and materials, what to do for their class display or how to reduce litter in the grounds. This will involve them in listening to their peers, realising that others have different ideas and coming to compromises. The National Curriculum also requires that children evaluate the success of their products and how well they achieved their aim, which can involve thinking about the effect of their actions on others and whether the costs justified the results. There is no point in such an evaluation if the activity has been closely teacher directed. Therefore, once the children start to plan and carry out their project, they need the opportunity to decide what action to take, choose materials and select equipment to use. This will inevitably mean that they make mistakes and have to contend with them, albeit under the guidance of the teacher.

By encouraging children to take an increasing role in the decision making, technology provides the opportunity to help children to become independent, thinking adults who are more likely to cope with problems and failure, and who regard difficult tasks more as challenges than barriers. By discussing decisions and evaluating their results they should also increasingly appreciate that their actions affect others and that they need to look for solutions that take into account everyone's interests, not just their own.

Developing tolerance and understanding of others

Cooperation, and sensitivity of the needs of others, is further fostered because technology requires that children work in teams to create different products and design organisational systems. This can only be realised by discussing and teaching collaborative and leadership skills, including helping the children to identify rules for behaviour and methods for optimising problem solving.

Primary children should not only consider their own needs and those of their peers, but should evaluate and create products for people from a variety of backgrounds and cultures. In order to devise appropriate products for parents, old age pensioners, individuals with disabilities and people of different cultures it is important to try to empathise with their requirements. When the children study homes, clothes, methods of food preparation, diets and religious buildings, if they understand that each product is a logical response to factors such as available materials, climate and social requirements they may be less inclined to criticise or dismiss the lifestyle of others as strange, and by doing so hopefully become more tolerant and understanding.

Awareness of conflicts of interest

Not only are junior-aged children expected to recognise the points of view of others and consider what it is like to be in another person's situation, they also need to appreciate that the range of criteria which must be used to make judgements about what is worth doing may be conflicting. On a mundane level it may not be possible to satisfy both the need to have a long-lasting pair of shoes and the requirement to spend very little money. More significantly there is the conflict between industries that produce many worthwhile products but may use irreplaceable resources or pollute the environment in some way.

Older junior children should have the opportunity to discuss a range of such conflicts between individuals or groups of people; between long-term and short-term interests; and between economic progress and the need to protect the Earth's limited resources. For example, children designing a zoo may have to choose between satisfying the demands of the public to see the animals clearly and the requirements of the animals for a suitable environment; between taking animals from their natural environment and the need to educate people to look after animals; and between conserving the ecological balance in natural environments and collecting rare animal species for breeding in zoos. Such technological activities can help children to identify such conflicts and sometimes accept that it may be impossible to find the ideal solution.

Throughout these activities children should become increasingly aware that their actions affect others, possibly detrimentally, and that every individual has a responsibility to look for creative compromises that protect and care for others in society and their environment.

Developing creativity and logic

Children need to develop creativity and skills of analysis and problem solving in all areas of the curriculum but the opportunity for their development is more obvious in design and technology, particularly as primary teachers are under considerable pressure to focus the children's attention on specific areas of knowledge and concepts in other subject areas in order to respond to all the requirements of the National Curriculum.

Designers have to be analytical, investigative and objective in order to clarify problems, to plan how to carry out a chosen solution and to calculate the possible future effects of their action. They need to have a good understanding of scientific physical relationships and how things work; be concerned with how people think and feel; and operate in an efficient organised way, communicating effectively in order to deliver a satisfactory product on time. However they also need to be imaginative: thinking of many solutions or methods; seeing connections between apparently unrelated things; being open to new ideas, attitudes and unconventional methods; experimenting with ideas and materials; and taking risks within sensible limits. Therefore design and technology gives opportunities to develop all these skills in a focused way and most importantly in a safe, guided environment.

In addition to these wider aims, design and technology links with and complements virtually every other discipline in the primary curriculum to the benefit of all subjects by showing how the children's studies are relevant and can be applied in 'real' situations.

FURTHER READING

National Curriculum Council (1990) *Technology in the National Curriculum* London: DES and Welsh Office/HMSO.
Shepard, T. (1990) *Education by Design: A Guide to Technology Across the Curriculum* Cheltenham: Stanley Thornes.

Chapter 2

Planning to develop skills and concepts

Design and technology describes a way of working in which pupils investigate a need or respond to an opportunity to make or modify something. They use their knowledge and understanding to devise a method or solution, carry it out practically and evaluate the end product and decisions taken during the process. These basic skills of identifying need, planning, making and evaluation should be progressively developed through the primary school.

IDENTIFYING NEEDS AND OPPORTUNITIES

Wherever there are people there will be problems requiring solutions. This may be helping an individual to cope better, such as enabling an arthritic person turn a tap, or making groups' or even societies' lives easier in some way. In order to see needs for themselves children should be involved in the day-to-day process of identifying problems and setting out to solve them. This may be achieved by the teacher sharing some of the classroom decisions and by enabling children to tackle open-ended tasks where problems will arise naturally. Alongside this involvement children also need to be helped to observe and evaluate their surroundings and manufactured products so that they can judge how successful they have been in achieving their purpose and by doing so will see further opportunities for designing improvements.

At present primary-school children have limited opportunities for learning to identify needs for themselves as many teachers usually decide what the children will do or make. However design and technology is less successful when teachers identify the problem and prescribe the route to a particular solution or end product, giving the pupils little scope to think and plan for themselves.[1]

Involving children in classroom decisions

If the children participate in some of the decisions about their undertakings they are more likely to develop an awareness and imagination for seeing needs and opportunities. Initially some teachers are very apprehensive about giving such freedom to the children, but those who do usually discover that the response is sensible, enthusiastic and frequently prompts far more excellent ideas than a single teacher can think of. Of course responsibility for the learning process is not handed over to the children, as a relevant context and guidance are essential to prompt the children's ideas and to ensure an appropriately balanced curriculum is provided. These contexts may arise from the day-to-day classroom management, a theme or project, survey or interview, visit, or talk from an outside speaker. For example, instead of the teacher deciding and setting up different role-play areas alone, the children could discuss and plan one of their own choice. They could be involved in thinking of ways to acquire different fabrics and how they can be effectively stored in the classroom in preparation for a science project on materials. Arrangements for a trip, a puppet play, a class or school party, or a class assembly can be shared with the children. In the last case, for example, the children could be involved in choosing or writing the material; planning and arranging the music; and collecting, designing and making the costumes.

Once the task has been decided, whether by the teacher or in conjunction with the children, an open-ended element where the method or detail of the end product is not too prescribed will provide other opportunities for identifying needs. In the past teachers have tended to protect children by providing only the materials they need and detailed instructions so that they are guaranteed a good end product, but this will restrict their opportunities to learn to choose appropriate tools and materials and solve further problems as they arise. The teacher's role is to focus and clarify the children's thinking, advise on the feasibility of suggestions, make resources available and provide guidance during the making process. Inevitably the children will sometimes suggest ideas that are inappropriate and the teacher should help the children to recognise this, as taking into account limitations and adjusting plans accordingly is a part of all design and technology projects. Children will also make mistakes which they will probably be able to correct themselves. However if the children are pursuing an inappropriate line of action or appear to be struggling the teacher will need to intervene to assist them to understand what has gone wrong and help them to think of a way to adjust their work. This process of error, review and adjustment will probably mean that some children will need more than one chance to produce an acceptable piece of work.

The teacher's confidence in enabling children to take an increased part in the decision process can be built up slowly by initially enabling the children to make suggestions within a fairly focused project. As the teacher and the children become more confident in discussing, negotiating and sharing decisions the projects can become increasingly more open-ended. For example many primary schools already celebrate or record special events such as Eid (Muslim), Divali (Hindu/Sikh), Christmas, a birthday, thanking a visitor or sending something to a sick friend. The teacher may decide that a card of some sort should be the end product but still wish to provide opportunities for identifying needs, evaluation, design and making. The children could look at different cards manufactured for similar events, consider what features they like and dislike, and discuss who is to receive the card and its acceptable content. Once the project has been discussed individuals can decide on the size of the card, materials, style, features to be depicted and then plan and produce their unique design accordingly.

On the other hand the same task could be presented to the children as a completely open problem. The children could brain-storm many ideas and discuss various possibilities such as making a card, a bunch of artificial flowers, a gift or traditional cakes and sweets. Groups of children could plan how to carry out different suggestions and list the equipment and materials needed. These could be reviewed, accepted, rejected or improved by the whole class. The children could then carry out the agreed tasks and finally evaluate the success of the activities.

In addition to the teacher sharing and encouraging the children to participate in projects initiated by the teacher it is also important to be alert to the children raising their own questions. In one school a group of children were worried that the hamster was being fed with inappropriate food or occasionally being let out of her cage. Having identified a problem and need for themselves the children were encouraged to solve it for themselves. They tried different unsuccessful ideas, including a poster which the hamster ate, organising patrols and an announcement in assembly, before finally rearranging the equipment in the classroom to get a safe location for the cage. In another class the children were keen to set up a fair experiment which entailed watering their seeds with the same amount of water each day. The children raised the problem of watering the seeds during weekends and subsequently experimented with different solutions to solve their own problem, one of which was to fix a water container above the plant so that a regular amount of water dripped into the pot.[2]

To further develop the children's ability to recognise needs for themselves, rather than have a problem identified for them by the teacher, they need to develop their ability to observe and describe existing artifacts and their surroundings in order to identify where improvements could be made that will benefit both themselves and others.

Observation and description

Reception children should be describing what they see around them in order to think how to improve, enhance or extend different features. They might describe their clothes and consider how well they suit the current weather conditions so that they can choose outfits for different climates. Descriptions or drawings following a visit to a nearby clothes shop may develop into setting up a role-play clothes shop in the classroom. A walk in a local park can stimulate discussion and lead to painting which can form the foundation for talking about and planning their ideal play area.

As children get older their observations should become more detailed. Their expertise will depend on the experiences provided by the teacher; questions which focus the children's attention on design features and their purpose; and opportunities for the children to articulate and discuss what they see. For example when describing a room the children can be asked to look at specific aspects including the size and shape of the room; the fixed features which cannot be easily changed such as the doors, window, built-in cupboards, electrical points and lighting; type of furnishings and flooring which can be varied if sufficient money is available; and decorations. Then each category can be evaluated with respect to the purpose of the room and whether it is liked or disliked by the children. Such detailed questioning will help the children to realise what to look at and give them ideas for improvements. Without this type of observational experience of existing products it is very difficult to design anything, such as their ideal room, from scratch.

The way the children's observations and evaluations are recorded should also become more varied and sophisticated and might include detailed written reports, annotated diagrams, models, graphs, tables, maps, sketches, videos with commentaries and so on.

Children asking questions and gathering information

The youngest primary children will usually be considering improvements that relate to their own needs and those of their class. As they get older they need to be able to ask questions to help them find out about other people's requirements. Initially the children may ask relatively unstructured questions such as asking a friend what materials and colours are liked so that they can draw/design customised party clothes. In order to discover more accurate information about the needs of others, who are less well known, the children should use more formal ways of gathering information which may involve designing questionnaires, conducting interviews and then analysing the data by using graphs or computer databases.

Thinking of the needs of others

As older children think of the needs of people from different generations, cultures and societies they need to learn to appreciate that such people can have very diverse points of view. A sensitivity and empathy for these groups may be achieved by using role play and by exploring the logical nature inherent in design. For example the children can investigate how religious or cultural demands, different climates, available materials and equipment, and the requirement of different types of jobs influence the design of clothes of all people including their own, so that they can design suitable garments for others. As the children explore the needs of others they should also be helped to realise that as different demands cannot always be reconciled the designer's task is to achieve an acceptable compromise.

PLANNING HOW TO CARRY OUT A PROPOSAL

Children are often reluctant to plan out their ideas, wanting to go straight to the making stage. This may well result in few alternatives being considered and the construction being abandoned as unfeasible. Consequently it is important that the children recognise the value of planning and attempt tasks, such as designing a garden for old people or a futuristic house, where the plan is the end product in its own right. Even very young children should have the opportunity to plan projects even if this stage is very short or rather informal. The expertise of older junior children can be extended by guiding them through a series of planning steps which they can see will help them create better products.

Plans and simple models in the early years

Very young children have limited experience of the properties of the materials they use so are often unable to anticipate how they can be used effectively. This experience will develop through trial and error during the making process. However it is important to introduce the idea and reasons for some preplanning. For example when several people are going to do a task together it is important to share ideas before starting. This might be done orally or in picture form. The projects could be a class display following the story of 'Little Red Riding-hood', a group model of their ideal playground, or a role-play post office for the classroom. In these cases the children can recognise the need to collect some information first, brain-storm ideas, and make small paintings, pictures or models which are then discussed and adapted before starting on the main project. The children will also be able to appreciate that preplanning helps the teacher to collect required materials. One reception class, having made imaginary animals out of a wide range of recycled and other constructional materials, were asked to design a home for their animal. The children drew their proposed homes and listed the materials they needed so that the teacher could collect the materials. When the homes were made on a subsequent day most children followed their plans with considerable accuracy.

Occasionally the children can also be encouraged to draw a diagram before making a model in Lego or other manufactured constructional material. In many cases this has the effect of making constructional play more effective, as the children are more ambitious in their projects; are able to share ideas with each other more; and learn more about the potential and limitations of the material when they try to carry out their plans. This use of constructional materials can later be extended to using manufactured kits to trial different ideas before using more permanent materials.

Developing a series of plans with older children

Older children should be able to develop more detailed plans but they too need to see this as a genuine part of the working process. A class of year 3–5 children were set the task of recording the visit of an Indian dancer. Some children were fascinated by the patterns of the dance; others were most impressed with the costume; and others by the hand movements. Consequently the range of techniques and activities they wanted to use varied considerably, including wanting to use printing techniques to record the repeating pattern; making clay models of the hand shapes; constructing a full-sized model using boxes; and producing small sculptures in wire and plaster. The children could appreciate that in this situation they had to plan their requirements in some detail in order to check whether there would be enough time to carry out their proposal, if the materials and tools were available or could be acquired and whether they understood all the skills necessary to do the proposal. In another project a group of children decided to produce a collage to decorate the classroom area by first tie-dyeing a piece of fabric and then adding further detail by printing and sewing. They too could see the necessity for drawing sketches of the final product, identifying the appearance of each stage and how to time it.

It is helpful for both the teacher and the children to have a series of steps outlined to remind them how to plan effectively. These stages are not entirely progressive and earlier stages are often revisited.

1. Clarifying the problem and identifying any requirements and limitations: One class of year 4–6 children were asked to design and make working toys, which used elastic bands, for reception-aged children. This involved asking questions such as: Who exactly is the product for? What time is available? What materials are available and at what cost? What tools and skills are needed? Do we have appropriate skills or have access to someone who can advise us? Will the product work and, if so, how? Is it expected to be attractive or primarily functional? What safety aspects must be considered? What are the possible consequences of making the product in environmental terms such as limiting waste of materials?

2. Brain-storming and collecting several possible ideas: The children collected as many ideas as possible for using elastic power which included making a paddle boat, a propeller-driven boat and aeroplane, a turning windmill, mobiles, moving cars and a catapult to fire an aeroplane or missile. The collection of ideas may also involve looking to see how others have solved the same problem in the past by looking in books and museums and by asking experts.

3. Researching feasible projects: Some ideas were rejected immediately as being unsuitable. After watching and talking to the reception children several different prototype models were made by trial and error. Then in a class session the models were reviewed in terms of safety, appearance and effective working and appeal to very young children. A few designs were finally chosen. Improvements were suggested and tested such as experimenting with the size of elastic band and the shape of boats and paddles. The reception children were asked about their favourite colours and pictures in order to choose suitable decorations and colours for the toys.

4. Details of construction, order and organisation: This led to the children thinking about how to organise making the final toys so that tasks were done in a logical order, responsibilities were shared appropriately and time was used sensibly and not wasted by waiting for glue or paint to dry.

Some of these stages were fairly undefined but they helped the children to think beyond the first reasonably feasible idea that they had and also helped them to see the benefit of using plans to learn from other people's problems, solutions and successes. At the end of the project the whole process was reviewed and the toys were evaluated by watching the younger children play with them.

MAKING OR CARRYING OUT THE DESIGN

The skills of using different media, tools and equipment are essential as a basis for design and technology. The children need to build up a range of techniques and should increasingly understand the potential and limitations of different materials. Without this they cannot make informed planning decisions on ways of solving problems and their achievements could be seriously hampered by a lack of technical skills. The school needs to identify the skills they want to introduce for handling all types of materials and decide at what stage these will be introduced to avoid teachers not building on earlier work or expecting too much of their pupils.

Some skills may be taught to individuals as the need arises as part of a technological project. On other occasions the main aim of an activity will be to focus on teaching a particular skill. The teacher needs a balance between providing activities that are teacher led, which are intended to teach children techniques or use of tools, and those that are more open-ended to enable the children to work through the complete design process.

Children in the early years are expected to use a variety of materials and equipment to make simple things. They need to know the characteristics of materials to make choices and are expected to be able to use simple tools, including basic woodwork tools and clay tools. Juniors should be able to choose appropriate tools and use them accurately having consideration about the quality and finish of the final product. They should consider problems of time and limits of resources and improvise accordingly.

difference between FPT & DMA - teacher's role.

Already in schools children frequently participate in art and craft activities which use a variety of artist media, recycled materials and textiles including sewing, knitting, fabric collage, dyeing and patterning fabric. Cooking tends to be carried out more in the early-years classrooms and this needs to be extended throughout the primary range. As equipment used for food should be kept for this purpose only it is probably worth keeping a central store of basic implements. However, as many recipes do not require heating or an oven is only needed at the end of preparation, activities can often be carried out in the classroom under indirect supervision of the teacher as long as the surfaces used for food preparation are clean. A small group of junior children can work independently while others carry out different activities. Indeed such an organisation means that the children have to take more responsibility in reading and following the recipe as the teacher may be busy with others. Children usually accept this arrangement as fair, as long as they realise that they will all have a turn eventually and get the chance to eat all the products! Of course an electric ring or small portable cooker in the classroom extends the possible scope of food products, although this needs much more supervision.

Many schools also need to increase the use of resistant materials (e.g. plastics and wood), teach the children to use the associated tools and then make them available for general use throughout the primary school. Basic experiences can be provided in the nursery and reception classes. For example early experiences using wood might involve using wood off-cuts and scraps for constructional play. The children can then be introduced to the problem of fixing their constructions permanently, initially by using glue and then with other techniques. These young children are then able to learn to use saws, hammers, nails, tacks and sandpaper to vary the way wood is fixed and shaped. [The advantage of teaching very young children to handle and care for such tools is that the children quickly identify them as normal provision and automatically consider their use in problem-solving situations. Important habits of safety are well established early and the children are far less likely to be silly with such tools if they are not a novelty.] Once the use of basic tools has been established the range and type of tools can be extended.

Many schools do not provide clay because they do not have a kiln and because constant replenishing of the clay is expensive. However it is not necessary to keep all clay products for posterity. The more frequently clay is available to the children the less they feel they need to keep their artifact, so as part of the evaluation process the children can be involved in deciding which pieces should be kept. Obviously a kiln is desirable but it is not essential. The few carefully chosen items can be dried and varnished or the help of a nearby secondary school with a kiln could be sought. Ideally clay should be always available for the children to choose to use as a means to express themselves or to solve a problem. If the children are taught how to prepare the clay and to store it properly they become more aware of the limitations, potential and processes in clay work and the pressure on the teachers to constantly monitor and prepare it is greatly reduced.

During most construction activities some noise and mess is inevitable as the children will usually be working in groups. By collaborating with adjoining classes or bases in open-plan schools, so that such noisy activities are synchronised, disturbance to other colleagues will be reduced. Mess can also be managed if the children are aware of the responsibility of using materials economically, without excessive waste, and of cleaning their working areas. If this expectation is developed from the reception and nursery years the children gain by having increased responsibility and the teacher is released to concentrate on skill development rather than organising and preparing materials in later classes.

EVALUATING WHAT HAS BEEN DONE

Children should evaluate their own work and existing artifacts, systems and environments.

Evaluation by children of their own work

Self-evaluation not only helps children to improve their own performance, it also provides opportunities for teachers to assess the children's progress. Early-years children need to describe to others what they did and how well they succeeded in their intentions by comparing the final outcome with what they intended to do. Junior children are expected to take into account not only their preferences but also how they met the needs of others. Primary teachers are rightly concerned to build on the skills and knowledge of the children and emphasise praise rather than criticism. Therefore the discussion of the success or otherwise of children's work needs to be handled sensitively. As self-evaluation is important the teacher can encourage individuals to comment on aspects they are pleased with and things they would do differently another time. However, in whole-class discussions and particularly when encouraging other children to evaluate their peers' products, positive features need to be emphasised by asking for comments on what they like about other children's work, and why. These points then can be used to suggest how the class as a whole would act if the task was repeated. It is often helpful to ask young children to choose two products that they particularly like, and give their reasons, as this allows an individual to choose their own work but also to look positively at the work of others.

Evaluation of manufactured artifacts, systems and environments

The early-years children on the whole will evaluate familiar products with respect to their own needs or those of their peers. They should be encouraged to say what the product is for, try to suggest how well it succeeds and describe to others what they like or dislike about it. As the children get older they should examine more complex and less familiar products, including those that were developed in the past and in other cultures with respect to the needs of others. Evaluating systems, artifacts and environments of other cultures has to be handled sensitively with the emphasis on what we can learn from other cultures and appreciating the reasons behind the use of different materials and arrangements. By the end of Key Stage 2 children should be considering the costs of products in the wide context of considering their impact on the environment. For example they could discuss use of CFCs as a possible cause of global warming, and use of non-renewable energy sources and scarce materials.

It is easier to evaluate several products of a similar type, as comparison helps children to identify the significant criteria for assessment. Initially the teacher may wish to suggest what criteria to consider and how to assess them. Testing is an important technique to analyse such aspects as the quality of the materials used and structural design. The children could test paper towels to find out which are more absorbent, or contrast the strength of carrier bags to discover the best buy. Information about people's preferences can be ascertained by interview and questionnaires or by recording behaviour. Favourite crisps could be discovered by tasting and interview sessions, asking for questionnaires to be completed or by recording numbers bought from the tuck shop. Cost is always an important factor in evaluating products, and when possible the children should have the opportunity to consider costs in respect to the quality of similar items. For example they could compare breakfast cereals to try to integrate the idea of taste with value for money.

As the children's ability to evaluate different products develops they can be helped to identify their own criteria for assessment and ways of judging them. The children need to be clear about who each item is for and its purpose in order to make a list of required features. For example when evaluating school chairs the children might suggest they they should be suitable for reception children, comfortable, washable, easy to move, stackable and visually attractive. A chair for a different situation would have other criteria. The size and shape of different chairs could be assessed by finding out the average measurements of the reception children to see how well the contours of different chairs conform and a poll could be taken to find out which chair designs and fabrics were preferred.

Evaluation frequently leads to other design and technology skills as it reveals areas of potential improvement which can form the starting point for further plans and making activities. Identifying a need, planning, making and evaluating is a cyclical process and it may be difficult to distinguish between the stages as each may be revisited several times in one project, as would occur if problems arise during construction forcing the children to see a need to make adjustments and replan. The different stages may also have different emphases depending on the project. In one, the planning process may be fairly minor and consist of oral discussions, but in another it may be the most important element. For example one class used a borrowed wheelchair to evaluate the design of their own school building and to suggest alterations to cater for children with physical disabilities. In this case, although they planned to widen doorways, provide lifts and build new ramps they were unable to carry out their suggestions.

It can be seen that the teacher needs to plan a wide variety of technological activities in order to develop the whole range of skills inherent in identifying needs, planning, making and evaluating and to give a range of contexts where these skills can be applied.

INCORPORATING DESIGN AND TECHNOLOGY IN THE PRIMARY CURRICULUM

When planning for design and technology the school needs to decide whether to have timetabled sessions or to incorporate the skills and concepts within other curricular areas. HMI reported in 1991 that most schools included design and technology as part of a topic or theme and it was often linked to science.[3] Such a cross-curricular approach has many advantages. It is already an established and familiar style of approach to primary teachers. The children are likely to appreciate the relevance of what they are doing, should see real opportunities for design and technology activities in context, and will have covered related background knowledge. For example, a successful bridge to carry a model car is more likely to be produced if the task is part of a wider project on the forces and structures of different bridges. In addition as the starting points should come from a wide range of subjects there is a better chance that the activities will appeal, at some time, to all children of different gender, race, culture and interest. This approach is also likely to be flexible enough to enable the timetable to be varied to respond when long sessions are required to develop a major project.

When design and technology is incorporated into other subjects it can be difficult to monitor the children's experiences and ensure that sufficient time is spent on the subject. Indeed HMI reported that, as whole-school plans and records of progress were rare in design and technology, there was a lack of continuity and clear lines of progression. This had the effect that Key Stage 2 classes paid too little attention to the development of skills learnt at Key Stage 1. Children were often engaged in activities that were too easy for them and were occasionally trying to use sophisticated equipment without the appropriate background experiences.[4]

By providing specialised lessons the teacher is able to ensure that enough time is spent developing skills and concepts, although a whole-school plan is still essential. However there is a risk that the teacher is continually forced into contrived pre-set problems which might lead to overimposed solutions and approaches that concentrate too much on artifacts, thus severely limiting the potential of design and technology. Some form of compromise is probably ideal where most of design and technology work is incorporated into other curricular areas, with some additional sessions focusing on specific skills related to handling a wide range of tools and materials, and others introducing concepts, such as those related to buying, selling, profit, markets and advertising, that are unique or particularly significant in design and technology.

It is essential to have a planned progression throughout the school which will obviously need to be based on levels of attainment and programmes of study outlined in the National Curriculum. In places these guidelines are rather vague, particularly in respect to the use of media and tools. In such cases the school needs to consider each category and decide what skills should be introduced and at what age. For example a list of techniques for decorating fabric (including tie-dye, screen printing, iron-on colour, batik, sewing or glueing fabrics, buttons, sequins and beads, and embroidery) could be drawn up and then discussions held to agree when to introduce them.

All whole-school plans need to be constantly reviewed, and this is particularly the case with design and technology as it is a new subject in the primary curriculum. As teachers' confidence and expertise develop they will become more aware of which skills and activities are most appropriate for different ages and which naturally lead on from others. In order to identify these, careful records are needed of what activities are tried and their relative success; which skills and concepts are introduced; and what difficulties were caused by lack of previous knowledge. As day-to-day plans are always being adjusted to take into account the children's needs and interests, such records will not only be useful to improve the long term plans but will also inform succeeding teachers about what the children have actually understood and experienced and what skills and concepts still need to be covered.

PRINCIPLES FOR DRAWING UP YEARLY PLANS

Once a progression of skills and concepts has been identified these need to be linked to other areas of the curriculum. It is essential to look at each year as a whole to ensure that all aspects of design and technology will be covered as some concepts fit in naturally with one topic but not others. When drawing up a year's plan it is helpful to check to see if the following can be included:

- Work on artifacts, systems and environments.
- Contexts from the home, school, recreation, community, and business and industry as appropriate to the age of the children.
- Evaluation of manufactured products from the children's own culture, from the past or other cultures, as well as making products.
- Examination of differences in personal taste and likes.

- Introduction of new skills in using textiles, construction materials and food.
- Opportunities to develop planning, making and evaluation skills during the term, although not necessarily all in one activity.
- At least one termly activity that allows the whole design cycle to be developed in one project.
- Work related to several areas of the curriculum.
- Introduction of new concepts such as those related to energy transfer and control.
- Development of collaborative group skills and organisation of people to do a task.

Teachers can brain-storm ways of including design and technology in their planned termly projects. These can be reviewed to see if any major aspect has been omitted or unnecessarily duplicated and additional sessions planned where gaps are identified. If an activity will unnecessarily duplicate skills or concepts or is at an inappropriate stage of the children's development, however interesting it may be, it should be excluded.

To illustrate this approach the example of a year 4 class is explored in detail.[5] In order to meet the requirements of the National Curriculum a school might decide that they will study five topics during the year, e.g. Weight and Gravity, the Vikings linked to Forces and Energy in Water, and Landscape Development combined with Rocks and Soils. Within this work there will be many opportunities to develop technology skills alongside most of the work, some of which are suggested in Table 2.1.

Reviewing the proposals in Table 2.1 it can be seen that a range of contexts are provided including recreation, school and industry. The children will have the opportunity to use different materials such as textiles, constructional materials and food in a variety of contexts. The school may feel it is unnecessary to use a database with two projects and so may omit one.

Such planning should not stop the children following up problems that they raise themselves and undertaking other activities arising as part of the minor projects that are always occurring in the primary school, such as making pop-up books for a book week. However by such long term planning a series of activities that will cover a balanced range of technology skills will be ensured.

Catering for all children

Encouraging equal opportunities in technology may meet with resistance from the children themselves as their attitudes to handling constructional materials, machines and ability to empathise with different people reflect deep-seated social patterns. These differences appear to be significantly related to gender rather than to cultural, linguistic and religious factors.

Table 2.1 Planning for design and technology: opportunities for integrating design and technology into projects planned for a year 4 class

Project	Possible Design and Technology Activities
Weight and Gravity	Recipes, to practically use and calculate different weights, to include some of different cultures. New cooking skills and techniques to be introduced. Discussion of how available ingredients, ways of preparing food, climate of country and religious requirements affect the design of cuisine. * Some cakes or biscuits to be prepared and advertised for sale at a profit during break. A spreadsheet package for sale of cakes and simple desktop publishing for advertising to be used. A survey of the most popular biscuit recorded on a database and analysed. Evaluate manufactured weighing machines and analyse them as systems. Design and make a machine to weigh letters to travel airmail.
Vikings & Forces and Energy in Water	Collection of typical examples of designs used by Vikings, decorate fabric using a variety of techniques, e.g. sewing, iron-on colour, and blocking using wax and dyeing. Constructing model boats which represent different periods using woodwork tools and balsa wood, and examining the design factors of boats in the past. * Design, plan and produce a boat to travel 1 metre under its own power.
Landscape Development & Rocks and Soils	Set up database to record information on rocks and soils. Children to organise a visit to a museum to see rock and fossil samples. Evaluate the environment of the museum. * Children to design and make their own display in the classroom for their samples. Considering how rocks and soils are used in an industrial system like the premixed concrete industry.

* Major projects

Differences in early play experiences lead to an imbalance in competence in handling constructional materials and in sensitivity to social interactions, as boys are more likely to have had experience with constructional toys or used tools and resistant materials, and girls are more likely to have been involved in role play. Not only are the children's experiences different, but they also frequently use the same material in very different ways. Boys tend to use constructional materials in more sophisticated ways, making greater use of it as a medium and exploiting its three-dimensional properties, whereas girls frequently make simple structures and use them as a means for social play.

It is important to encourage children to participate in a wide range of activities from the early years. Cross-domain play can be encouraged by an adult being involved in the activity, particularly when a teacher of the same gender as the children can be involved. Boys, for example, are more likely to engage in what they perceive to be a 'girls' activity if a male member of staff is present. Constructional activities in the early years should also be regarded as 'work' with the children encouraged to finish their model as any other piece of work and it is important to have high expectations of all children and to praise accordingly. It has been found that when teachers try to encourage more constructional activity girls employ a variety of avoidance strategies, including making something very basic which requires minimal effort and then showing it to an adult who often gives excessive praise to encourage them to use the material more. Unfortunately this is more likely to have the effect of reinforcing under-achievement.[6] Having completed a model the children should be encouraged to play with it or make up a story to go with it. Alternatively they could be asked to make a model to illustrate a story previously read to the class, such as making a castle for the 'Sleeping Beauty', or be asked to construct a model for a particular purpose like a lifeboat to save a child and dog cut off by the tide or a milk float to deliver milk to a nursery. By providing a social context children are helped to move away from stereotypical play and introduced to the idea of making products for a specific need, and this should encourage girls to identify with the subject as their interest is usually captured when related to home life or a social need.

It is also important that design and technology projects throughout the primary school should require a broad range of materials, including handling a variety of fabrics and plastics such as corriflute, and be related to all areas of the curriculum. By providing a very wide range of activities, such as preparation of food, making moving vehicles, designing clothes, building a school pond, preparing a dance, arranging outings and improving the sale of biscuits at play-time, all children should find something that appeals to them. It needs to be made clear to the children that all these activities are technology so that they do not have a narrow view of the subject.

The very practical nature of design and technology should enable children with learning difficulties and those who speak English as their second language to take a full part. This can be assisted by emphasising oral communications and by encouraging a wide range of recording methods including collaborative writing, photographic records, taped reports and pictorial representations.[7] Children with physical disabilities may have difficulties with some of the tools and processes, but team work and the involvement of support staff should enable them to participate in most activities. It is important that the children should be in charge of their work and be expected to ask for help when they feel it is necessary and to specify the assistance they require. It may also be possible to adapt equipment to assist them. Their peers could even try to solve their particular needs as technological projects.

Those children from different cultures must also feel that their contribution is valued by inviting them to contribute their special knowledge whenever possible. For example when evaluating or designing cooking utensils, jewellery, styles of dress or games, those of other cultures can be included.[8] Using the children's own experiences from different cultures makes technology relevant to them, increases their self-esteem and enriches the experiences of their peers making the requirement of the technology curriculum to study different societies and cultures more immediate and personal.

NOTES AND REFERENCES

1 HMI (1991) *Aspects of Primary Education: The Teaching and Learning of Design and Technology* London: HMSO, paras 16 and 32.
2 Greeves, J. (1990) 'The root of the problem' *Questions* Vol. 2, Issue 9, pp. 4–7 explores several ways of watering plants over a holiday period.
3 HMI, op. cit., para. 14.
4 Ibid., paras. 14, 21 and 23.
5 Several examples of a yearly plan are given in the Non-Statutory Guidance for Design and Technology. SEAC (1992) *Children's Work Assessed: Design and Technology and Information Technology* London: HMSO contains excellent examples of projects that have been integrated with science.
6 Browne, N. and Ross, C. (1991) 'Girls' stuff, boys stuff': Young children talking and playing in Browne, N. (ed.) *Science and Technology in the Early Years* Milton Keynes: Open University Press.
7 Mount, H. and Ackerman, D. (1991) *Technology for All* London: David Fulton reports on how a group of 7–10-year-old children with severe learning difficulties tackled a variety of projects based on a modern Noah's Ark story.
8 Barnfield, M., Comber, M., Dyble, L., Farmer, M., Hagues, S., Hughes, S., Martin, T., McFarlane, C. and Moore, A. (1991) *Why on Earth? An Approach to Science with a Global Dimension at Key Stage 2* Birmingham: Development Education Centre includes useful material on the themes of shelter, food preservation and energy.

FURTHER READING

Bindon, A. and Cole, P. (1991) *Teaching Design and Technology in the Primary Classroom* Glasgow: Blackie.
National Curriculum Council (1990) *Non-Statutory Guidance: Design and Technology Capability* York: NCC.
Smail, B. (1985) Chapter 4 Organizing the curriculum to fit girls' interests *Girl-Friendly Science: Avoiding Sex Bias in the Curriculum* York: Longman.

Chapter 3

Cross-curricular links

The skills of identifying needs, planning, making and evaluating need to be placed in appropriate contexts. These can be provided by linking design and technology activities with other subjects. In return those subjects can be enhanced by the children appreciating their application in 'real' situations, and teachers are given further opportunities to introduce and establish complex concepts and to assess the children's understanding. During technology activities the use of English and mathematics arises naturally, and as children are usually very motivated by its very practical nature the quality of their oral and written language, standard of measurement and calculations tends to be high.

Most people also readily recognise that science is very closely related to technology, as the latter is concerned with developing and applying scientific principles to meet human need. Once need has been identified a review and application of the relevant science ideas will improve the final product. When designing a bridge, if the children are encouraged to recall the forces that are likely to affect the bridge, the properties of the materials they have available and the methods for strengthening those materials, they are more likely to produce an efficient and economic design.

Such technological activities also enrich scientific learning by demonstrating its relevance. The properties of sound can be applied to explain how different musical instruments work and to evaluate their effectiveness; investigations into mixing the primary colours of light can be linked to theatre lighting; and principles of stability can be used to explain the necessity for having very wide bases on a baby's high chair. As the children later use these principles to make their own instruments, devise different theatre sets and evaluate furniture design, opportunities will arise to assess their scientific understanding.

Methods of science investigation can also be practised in technology as part of evaluating and planning technological products. One class who evaluated manufactured mugs, with a view to designing their own mug, carried out tests which included assessing their stability and heat retention. On another occasion, when designing clothes to take on a holiday to Pakistan, the children carried out tests on different fabrics to ascertain which would be cool and did not crease or stain easily. Such activities entail children identifying variables, carrying out fair tests, making predictions and suggesting hypotheses in order to distinguish the factors that are likely to improve a proposal as well as to discover which materials and designs are best.

Although the links with the core subjects are especially important, the possible wide scope of design and technology means that there is potential for incorporating technology into every subject in the primary curriculum. For example geography includes the study of the way people use their environment to provide for economic and social needs. This relates directly to the technological approach of studying an environment both to evaluate its success in achieving its purpose and to suggest appropriate improvements. Studies of environments can also give purpose for and practice in interpreting and drawing maps and plans, as these are needed to record, analyse and communicate observations and proposed adaptations.

Any historical topic which relates to domestic, industrial or social development, or to human interaction with the environment, must by its very nature have technological connections.[1] Rather than just accepting products of the past as quaint – interesting but not really relevant to life today – a technological approach helps children to appreciate how and why things have come to be the way they are at present and helps them to predict consequences of future change. Inventions and innovations were usually based on existing products and were influenced by available materials, manufacturing processes, labour, finance and current fashion, in the same way as they are today. For example early motor cars looked like the horse-drawn carriages they replaced and relied on traditional carriage-making skills. Early electric cookers were restricted to cast iron and appeared very like the older traditional solid-fuel stoves.[2]

By examining pictures and actual examples of products at different stages of development the children can suggest why they were invented and changed and perhaps suggest improvements for modern examples. They will usually find many stages with fairly minor changes, some more successful than others, that attempt to solve problems with the product or that apply innovations made in other fields, as in the case of the development of telephones and bicycles (Chapter 6). It may also be possible to give the children the challenge of solving similar problems as those of past designers, such as working out how to build a straight road like the Romans or making a Victorian magic lantern. Such tasks can give a stimulating context for technology as well as helping the children to empathise with the people of those times.

Not only can technology be effectively linked with all the foundation subjects but there are also close links with the cross-curricular themes. The theme of 'Environmental Education' involves using the environment as a resource to provide knowledge and to develop values and attitudes which lead to positive action to find solutions to environmental problems and to find ways of ensuring care for the environment now and in the future, taking into account that there are conflicting interests and different cultural perspectives.[3] This is entirely compatible with the approach adopted by technology, which usually starts by studying the existing environment with a view to identifying its potential and limitations, and then suggesting ways of improving and developing it in some way. Similarly, many of the aims for 'Education for Economic and Industrial Understanding' overlap with those of design and technology. Both require that children appreciate how consumers and producers relate to each other; that work places are organised in different ways; and how goods are produced, distributed and sold in different enterprises in their community.[4] Many of the attitudes required in the Citizenship theme,[5] including respect for others' opinions, ability to work with others and approaches to resolve conflict, should also be developed as part of design and technology. In addition, the component on work, employment and leisure can be developed when studying technological products in the contexts of recreation, business and industry.

It is the intention of this chapter to explore the relationships between technology and English, drama, mathematics and art in particular. Later chapters consider links with other areas of the curriculum.

ENGLISH

As language needs to be used in different ways within a wide range of situations, most curricular subjects can be regarded as a context for the development of English. There are some rich and exciting possibilities within design and technology in particular. Design-related activities promote language development through listening, speaking, reading and writing, and language work can offer opportunities for the development of design and technology.

By children working in groups, design and technology projects generate discussion and debate. They are motivated to listen to others, to articulate their ideas and to discuss possible solutions. The children's reading skills are enhanced as they consult books to further their proposals, and as they need to read such books selectively in order to choose appropriate information they can apply different information-retrieval strategies for genuine purposes. A wide variety of language activities develops naturally, such as oral or written creative pieces to produce a book or to imagine the effect of a course of action after the implementation of a technological solution; descriptive and analytical language as part of the evaluation of manufactured products; and sequenced explanations of how a device works.

Establishing a class radio station, for example, can embrace many different language skills in an exciting and stimulating context. In one class the children initially identified the advantages of a local radio station compared with a national one, such as being able to cover local events, personalities, weather and traffic reports. The children had a competition to choose an appropriate name and logo, and then different groups decided where they would site their station in the locality and set up a role-play area. They decided on the staffing, which included writing job descriptions to cover hours, pay, holiday, experience and qualifications required, and then produced job advertisements. The children wrote letters of application and had the experiences of interviewing candidates. The class discussed funding of the station through advertising and subsidies and ran an advertising campaign with official opening, leaflets, T-shirts and other goods. A range of programmes was produced, including designing and producing a play, presenting a selection of music, news reports and sports commentary.[6]

Furthermore, activities which are primarily language-based help to develop technological skills. A story enables new, perhaps unfamiliar situations to be introduced to the children within a secure framework, or a problem can be set in the context of a story so that the children can appreciate its relevance. Storytelling encourages the development of imagination, prediction and imaging necessary in technological activities. The children's own stories can be drafted and redrafted with a view to presenting them to a specific audience. Manufactured books can be evaluated in terms of construction, content, plot and structure, and presentation (including type of print, chapter headings, diagrams, illustrations and language level) which might be followed by the children producing their own books.

Stories as a context for technology projects

Stories can be used by the teacher to prompt the children to solve a problem raised in the narrative. In the story 'Old Bear'[7] toys in a child's bedroom want to rescue a teddy bear which has been put in an overhead loft. The toys try different methods, such as building a tower of bricks and bouncing on a bed in order to reach the loft. Once the children understand the aim of the story they are usually keen to suggest ways of saving the bear, for instance making a parachute for it that can be designed and tested and making a trampoline for bear to jump onto. A few of these ideas can lead to a making activity, but many are satisfactorily explored by discussion.

The story of the 'Three Bears' inspired a class of nursery and reception children to make a collection of teddy bears of widely differing sizes. The children were encouraged to build furniture to cater for each specific bear. Other projects might include designing and making a security system with alarms and traps to protect the 'Three Little Pigs' from the wolf; creating a device to get the 'Run-away Pancake' across the river without falling into the jaws of the fox; and making a model of the 'Iron Man'[8] with moving joints and hinges, flashing eyes and a magnet or electromagnet to sort iron from other metals in his hand. The story itself may act as the inspiration to present a play with props, costumes and scenery being designed and produced by the children, and accompanying music could be made to retell the narrative.

Two other stories that have been found to be particularly useful are 'The Lighthouse Keeper's Lunch'[9] and 'Dear Zoo'.[10] In the former story the lighthouse keeper's lunch is delivered by a pulley system from his home on the cliffs to the top of the lighthouse. Unfortunately it is regularly devoured by seagulls on the way. The seagulls are finally discouraged by adding mustard to the sandwiches. In one class the story led to children questioning how a pulley system works and building one to cross the classroom and building a working lighthouse with flashing light. In 'Dear Zoo', a progression of animals is sent to a child who has requested a pet. These arrive in different types of crates and containers suitable for the animal to be carried. This story led to children building travelling cages for different animals that moved, and were designed to cater for the needs of the animals and the safety of the keepers. In one particular class they continued their interest prompted by the book to design and make a model zoo. They researched the needs of each animal to produce suitable environments. It also led the children to question whether animals should be kept in zoos at all.

Books and stories which have characters with a special set of characteristics can be used to prompt children about the particular needs of others. The children could be asked to design a fireproof home for a dragon they have just read about. They will need to consider what the dragon eats, where it sleeps and what sort of environment it feels comfortable in. Characters such as Tom Thumb, who is a normal healthy boy but is only a few centimetres high, or Mrs Pepperpot,[11] who is sometimes an ordinary lady but without notice can shrink to the size of a kitchen pepperpot and just as suddenly revert to her normal size, can prompt the children to think of clothes or equipment the characters might need and what design of home might suit them best. The children could create and perhaps make a model of an imaginary character with particular characteristics and then consider its lifestyle, eating habits, clothes, language, home, property and means of transport.[12] This visualising can then be used to start a story about events in the individual's life. In this case a primarily technological activity can develop into creative story-telling or writing.

As the children mature less unusual characters can be discussed. By acting out a story the children can put themselves in the place of the character and gain first-hand experience of how that individual might think and act. Additional problems and events can be proposed for the character to explore how he or she might respond, or the story can be developed so that the children can consider the consequences of an action. The children might take on the role of Cinderella's ugly sisters who need to abandon the quest for husbands to find a job or think of other ways of finding suitable suitors, perhaps by improving their appearance or advertising for their own fairy godmothers. Such drama provides a good foundation for more in-depth role play around specific technological issues, such as taking on the parts of people in dispute over developing a housing estate in an area of natural beauty.

Not all books and stories lend themselves to technological activities and it is important to avoid contriving a link that may demean children's natural approach of raising questions or falsely give the message that literature is not to be valued in any other way. It is essential that the children feel that the story genuinely prompts questioning and problem solving. Additionally a reliance on published stories alone would be limiting, as oral story-telling by the children does much to develop imaginative and creative thinking.

Developing creative and logical thinking through oral story-telling

Children need to be able to picture and describe imaginary articles and situations so that they can suggest different ways of responding to problems and appreciate the potential for developing or even creating new artifacts, systems or environments. If a story is told without any accompanying pictures it can be used as a starting point for visual interpretation. The children might express their interpretation of an object, the environment or characters in words, pictures, charts, maps, time lines or three-dimensional models. For example, in the story 'Goldilocks and the Three Bears', the flow of the story might be interrupted just as Goldilocks picks up the 'great, big, enormous spoon' and the listeners asked to share their mental picture of that spoon, how it looked, what it felt like, what it was made of. Usually many different imaginary spoons are described: some plain, some with intricate patterning, a 'loving' spoon, a red plastic one.[13] The actual range of responses can help children to realise there is no right answer. This ability to bring a very individual image to mind can readily be transferred from the story context to making an object or designing an environment.

In order to solve problems and think of ways to satisfy needs, several imaginative ideas need to be generated. These are then reviewed and analysed to identify which one can be developed further, bearing in mind the possible consequences of its implementation. Oral story-telling helps to develop each of these aspects. The process of narrating requires creativity and involves techniques of questioning, prediction and hypothesising. Shared story-telling involves careful listening, prediction, speculation and an understanding of causality which includes using the evidence of the story as it develops. Children are able to listen to others, which often stimulates their ideas and gives them more freedom to change and explore an idea and its consequences without the laborious process of writing. Being able to formulate a logical story will help the children to develop the ability to forecast the possible consequences of technological products.

Initially children need to develop their confidence and skills in oral story-telling by perhaps taking turns in retelling a very familiar tale. Once the children feel at ease telling stories aloud they can be encouraged to vary and develop them. For example, in group- or class-shared oral story-telling sessions, the story-teller can be interrupted and asked to change a significant element of the story, such as who the Three Little Pigs met on the road or what Jack took in exchange for the cow. The teller is then forced to incorporate and account for the consequences and through such small changes the story can develop in unexpected and often amusing ways. Even quite young children can surprise their teachers by their inventiveness and attention to detail.

Stories consist of a pattern of events which can be used to provide a structure for imaginative development. One class had been reading 'The Run-away Pancake' which is a traditional story in which a newly made pancake runs away only to be chased by the woman who made it, her children, a dog, a cow and a horse who all want to eat it, only to be tricked and eaten by a fox who pretended to help it to cross a river safely. Once the children understood the pattern of the story they suggested run-away chapattis, cakes and biscuits which were chased by various animals and people, including in one case the whole staff of a school, to meet their end in different ways including a jungle and road accident. The stories created by the children can be used to form the basis for the design and production of their own books. After all many published stories use very similar techniques to change well-known stories in some way.

Such activities increase the children's confidence in their own creative abilities. Many children hold the view that they lack imagination but given a structure and a context they can discover considerable talents in this area. It is also important to recognise that the story changes demand logical as well as creative responses, helping to develop the children's abilities to predict and extrapolate, essential skills if they are to anticipate the possible effects of technological change.

Some published stories also lend themselves to developing prediction. For example many lift-the-flap books can be used to encourage the children to predict what they will see when the flap is lifted. Children are reluctant to make predictions in mathematics and science because they do not want to be wrong. This situation is one occasion when 'incorrect' predictions can be seen by the children to be a positive advantage. They are in effect creating their own version which could again form the basis for the production of a new story or book.

Alternatively a technological problem-solving activity can be incorporated into creative writing. This approach entails providing at least the beginning of a story and a problem. This might include the hero of the story being faced with the problem of how to escape from a sinking island, surviving on an uninhabited island or needing to traverse a shark-infested river.[14] Smith[15] suggested to a group of children that they should imagine that they were secret agents who had to reconnect an electrical circuit to a lighthouse in order to flash a message. The wire had broken in a well which was rapidly filling as the tide was rising. The children were given various materials in order to work out a way of overcoming the problem. On completion of the activity the children wrote the account of their mission from the point of view of the secret agent.

It can be seen that the potentially close relationship between English and technology can be exploited to enhance the learning in both subjects.

Books as technological products

Books can be considered as artifacts as well as a source of inspiration to raise technological problems.[16] Instead of the class book of the children's stories or topic work being collated by an adult, the children could be involved. After completing various stories an early years' class was asked to produce a class book to go into the library. They had to decide on the order of work, what illustrations should be included, how many pages of folded sugar paper they would need and how to fix the pages together. The children sewed the pages together and added a decorated cardboard cover. This multicultural class also decided to make a tape of each story, in the correct order, in both English and the home language of the writer which was then placed in a pocket in the front cover.

A class of junior children might make books for younger children which could involve them in carrying out a survey to discover what stories were particularly enjoyed, the type of illustration and print most understood, and what size of book small children liked to handle. On the other hand, older children might review real newspapers and magazines with regard to their style and intended readership with the intention of designing and producing a publication for parents. Producing multiple copies might then lead to decisions on whether handwriting, typing or word-processing is most appropriate, and how the material can be duplicated cheaply and marketed effectively.

Manufactured books can also be evaluated with a view to designing and making similar ones, particularly those with moving parts. The moving parts can be very simple, such as lift-the-flap, pulling a lever and turning a wheel to change a picture, or can involve complex pop-up designs.[17] These books can be designed and made even by very young children. For example in a year 1 class the children initially evaluated manufactured pop-up books to consider their advantages and disadvantages. The children not surprisingly wanted to make one but had none of the basic techniques necessary. As the children all spoke English as their second language, the skill was taught in conjunction with work on opposites. The children were shown how to make a simple lift-the-flap using two sheets of paper glued together. Small groups then thought of pairs of opposites and decided how they could be illustrated. One idea was to have a tall lady outside a door which lifted up to reveal a short woman. Each child produced one page, all of which were then attached together to make a book. Once the children understood how to make this simple mechanism a second one was introduced using the story 'Maisy Goes Swimming'[18] as a model. The children made a similar one showing the stages of getting ready for PE which helped them use different words and think about sequences. When the children were older they were able to draw on these early experiences to make more independent creations.

ROLE PLAY AND DRAMA

With the pressure on schools to plan long-term to satisfy all the requirements of the National Curriculum, and the need to collect resources before most activities, it is not always possible to respond to the children's ideas to solve a perceived need that arises by chance in the school or community. Teachers frequently require to find ways of providing opportunities for children to identify needs within a planned framework. This can, in part, be achieved through drama which will also provide the class with practical experiences, in a safe fictional form, relating to different communities, industries and locations that children cannot normally experience in the classroom.

Several fundamental skills that are needed in technology can be developed through drama. Children are given the chance to visualise imaginary situations with a view to suggesting practical changes and to explaining to others why they are appropriate. Drama also gives the children opportunities to empathise with other people. By acting out a given situation they can recognise different people's needs in present-day situations, in historical contexts and in different societies in the world. The children are able to explore some of the social, ethical, political and economic issues that arise from technological innovation and they can actually experience some of the emotions generated by conflicting views or concerns about difficult situations.

The children are also assisted to develop group cooperative skills as the drama enables the children to move away from their expected classroom personae and roles within a simplified structure. If the children have been asked to take on the roles of adults from an estate worried about a factory which is discharging large amounts of smoke, the children's interactions are guided by the drama rather than their usual classroom relationships. In addition, children who might be identified by their peers as having intellectual difficulties or are expected to be uninterested or disruptive can take on new roles which may cause their peers to look on them in a new light.

Contexts for raising problems

The drama or role play can provide a context for raising problems which stimulate the children's interest and by acting out the roles helps them to explore the situation in depth. The children might be asked to be relief workers helping to provide water for a Third World community who need to design methods of digging wells, raising water and purifying it. Their designs can then be made in model form as a way of testing their effectiveness.

As part of a project on rocks a class might imagine that they are geologists who have to explore a newly discovered cave system and analyse the rock samples collected.[19] They could prepare for their expedition by making equipment such as head gear and lighting devices. Once the equipment has been prepared the children can act out their entrance to the cave system, crawling through small tunnels, squeezing through narrow fissures, negotiating narrow and slippery ledges and entering huge caverns with small pools reflecting stalactites and stalagmites to collect imaginary rock samples and map the caves. The drama could then be supported by science work on actual rock samples. Subsequent drama could include dealing with members of the public who want to see the caves but may damage the rocks. This can be followed by drawing up plans for the imaginary site to cater for visitors, to include areas of limited and safe access, parking facilities, café and museum. An actual 'museum display' can also be designed and constructed using rock samples and information from their own science experiments.

In such role play, active participation by the teacher is often helpful to add additional elements or extend the view of the children. In this example the teacher could act as the director of a company which has commissioned the geological firm to survey the area and which wants particular tests carried out; as a belligerent member of the public who wants to break off a stalactite to take home; or as a local farmer asking for advice about the best way to treat the soil.

Another way of introducing an imaginary locality in order to give a framework for technological problems is to postulate a transport accident which leaves the children in a relatively hostile environment where they have to survive until rescue or escape. For example one class imagined that they were in an aeroplane going to the West Indies to visit relatives when the teacher in the role of pilot announced that the aircraft was in difficulties and would be making a crash landing. The children were encouraged to act out their responses and feelings. The aeroplane crashed into a tropical forest area. After the escape of some of the passengers (i.e. the children) the aeroplane blew up killing the pilot. The children now had to survive using the limited amount of equipment, indicated by drawings on cards, which were found in the wreckage and forest. Following the drama the children were encouraged to identify their immediate and long-term needs for food, water, shelter, security and rescue based on previous class work on tropical rainforests. The children identified specific problems to solve, such as catching and cooking food, deciding what food was safe to eat, obtaining safe water to drink by filtering and then boiling it, making a shelter, frightening off marauding animals, inventing a signalling device and designing a vehicle for escape, either through the air using a hot-air balloon made of thin cloth sewn together or by boats of various designs to travel down-river with crocodile protection devices. Some of these were only drawn as plans; in other cases it was possible to make models. Similar situations include being abandoned on a desert island, in an arctic environment, or on another planet after an accident in space.

Drama can enable children to appreciate that the apparent solution of a technological problem can create other difficulties or be unacceptable in some way. Kitson[20] reports on a drama where the children were asked to imagine a market gardener who was concerned about his crops being destroyed by birds and rabbits. The children took on the roles of trainee journalists who interviewed the gardener (teacher in role) in order to produce an article for their paper. The children were encouraged to think about the gardener's feelings so that they could draw out his views effectively. This led to the children producing ideas for traps to get rid of the pests. At this point an opposing viewpoint was introduced in the form of a tramp who had befriended the animals and a group of animal protesters. The children were paired so that a 'journalist' could interview a 'protester' for a second article. The children could now see that a killing trap might not be an acceptable solution so they tried to devise methods of removing the animals without harming them.

Exploring industrial concerns

In role-play activities, situations can be set up to simulate industrial concerns such as setting up a paint business. The children could be shown paint-shade cards produced by existing paint companies, and small groups of children be asked to produce their own. The groups can be asked to name their company and invent new paints with appropriately exotic labels. They can be reminded that each invention should be recorded in detail and in such a way that the paint can be reproduced and made in bulk. The companies might then advertise their product by writing to imaginary DIY outlets, producing newspaper adverts and producing radio (tapes) or even TV (videos) adverts which explain that their product has special properties such as ability to withstand hot steam or being easy to clean and apply.

Drama is a very effective way of helping the children to understand conflicts between industrial activity and protecting the environment, and the conflicts between individual need compared to the requirements of a whole community. Littledyke outlines a project for upper juniors in which a fictional village was postulated and the children took on roles of the inhabitants.[21] The pupils formed family groups of four or five and decided on their names, ages, relationships and employment. All were adults so that they had responsibility during the drama. In order to help establish the roles the children acted out short scenes of family life such as the evening meal and they gave a short account of their family history. The imaginary community was set in a location where the environmental quality of life was high but many people experienced problems of unemployment. The village was named and plans drawn up. As the children developed their drama the problems associated with unemployment were drawn out by acting out scenes in a pub and meetings in shops. Once the children had established their role, plans to build an agrochemical, reprocessing plant were then introduced by the teacher in role as a villager starting a rumour. As the children acted out their response, in role, to this news, the benefits of more jobs and increased opportunities for local business compared to the possible damage and disturbance to the village were identified.

In another school, as part of a science project on building materials and the work of the concrete industry, a class of 10–11 year olds were asked to act out the views and feelings of farmers, house owners and industrialists in the imaginary situation of a concrete industry wishing to develop land to extract gravel.[22] The 'farmers and farm workers' were concerned about the loss of their jobs and in some cases their houses, and the 'villagers' worried about the increased dust and noise and the danger of lorries that carried gravel constantly driving through their village. The three groups negotiated compromises which included offering farm workers jobs in the concrete industry, hosing down the extraction plant on very dry windy days to reduce dust, building a new road around the village for lorries and agreeing to restore the gravel pits after extraction for recreation. Finally groups of children produced plans to renovate two imaginary disused gravel pits to include activities such as skiing, sailing, fishing, cafés, car parks and quiet areas for wild life.

Historical contexts

Technological problems that have arisen in the past can be presented and given a context through role play which will also enhance historical understanding. This was done in one project where the teacher took the role of a Roman general required to establish a presence in Britain.[23] After discussion with his soldiers (the children) they identified what they needed to do to improve their basic fortified wood enclosure. The children decided to build a watch tower for increased protection from the unfriendly Britons which was designed and a model made. The 'general' then gave different tasks which included surveying an area of land to build a road in order to establish a supply route. The children were asked to find a way of drawing a profile of the land level, finding a method to assess the hardness of the ground and devising an instrument for sighting straight lines. Once the children had contrived their own solution to the problems it was possible to show them the machines used by the Romans such as the chorobate, a type of early spirit level, which was a heavy table with a trough of water in it which could be moved to keep the water level.

Social conflict which often occurred as a response to technological change can also be explored though drama. The children might act out the roles of families in the early 1800s engaged in the home knitwear trade when faced with the introduction of new machines. The teacher pretending to be a Luddite could secretly visit the families, giving information about minimum rates that should be paid and how other workers have put pressure on the hosiers by attacking or threatening the new machines. In order to explore the view of the hosiers the children can also act out a meeting of local hosiers to decide what to do about the unrest in their area: whether to buy off the workers, collect the machines in one place to protect them or threaten the work force by calling in the militia. The final part of the drama might be the villages having a meeting to decide what action to take.[24]

When studying the effect of transport changes after World War Two, the children could take the part of a family living on a canal narrow boat where the father is away fighting in the war. Such a drama might start with the children acting out their lives on the boats such as caring for the horse, loading coal, decorating cans and bowls, preparing food in the confined space of the cabin, fishing, etc. The return of the father gives the opportunity for everyone to share their experience of the war and might be an occasion for a party. However the celebrations are marred when the father tells his listeners that he wants to sell the horse and boat to buy a lorry. He feels the family will be better living in a 'proper' home where the children will be able to go to school now that schooling is compulsory, and the lorry will be more efficient and less trouble than the horse. Obviously the news about the lorry will not be welcome to everyone and the final part of the drama might end on a vote to decide the outcome.

In many of these dramas a structure can be identified which can be used as a model for other ideas. The first phase entails establishing the life and activity of people who live in a particular situation such as the factory, market or village. A threat to this situation is introduced, such as the open-air market traders being served with notices of closure to make way for a hypermarket. The final stage involves the children trying to find some solution to this situation.[25]

MATHEMATICS

'In theory, it is possible to grasp any aspect of mathematics. (For example the four rules of number.) But it is only when we have to apply mathematics to meet a real need that we can fully understand its relation to the world around us. Design provides an ideal setting for this.'[26] The mathematics National Curriculum requires that children use number, measures, shape and space and handle data in practical tasks and real-life problems, all of which can arise naturally within technological activities.

The design of many artifacts and environments requires the manipulation of basic number and measurement which should be geared to the mathematical ability of the children concerned. Reception children making a swing for a doll might match the size of the doll with pieces of wood before cutting, whereas older children designing a milk crate in plastic corrugated sheet (corriflute) can be expected to accurately measure the items to be carried, produce a prototype in cardboard and then take careful measurements in order to produce the final product.

Many goods have to take human dimensions into account. If the children are evaluating the furniture in the school they will need to take measurements of the range of body sizes and shapes in each class to ascertain whether the basic shapes are satisfactory. On the other hand the children might be evaluating or designing off-the-peg clothes which have to cater for average sizes. By comparing the children's actual age with the size they need to buy, they could consider whether clothes are reasonably labelled. They might then take measurements of children in a particular age group with a view to suggesting measurements for their own range of clothes.

The study of a business enterprise, such as a shop, involves handling and calculating money at all levels of ability and age, from recognising coins and carrying out simple calculations to computing production costs, profit and interest on loans. Surveys, requiring the manipulation, diagrammatic representation and interpretation of data, can be used to assess potential customer need and the effect of advertising campaigns, and to identify which products are most successful. The evaluation and design of packaging could include the children investigating different nets in order to make packages that use cardboard economically and how three-dimensional objects can be fitted or stacked together to make optimum use of available space. Studying the organisation of workers in an enterprise might incorporate considering opening hours, timetabling, and issues of paying basic wages and overtime.

Making food items can require the children to read recipes, adjust the amounts to cater for a different number of people, buy the ingredients, weigh them out, calculate the cost of each item and perhaps work out an appropriate profit. Over a year groups of children from a class of 8–9 year olds took turns to choose and make different food dishes from around the world. Each group was asked to choose a recipe from a particular country and then plan, buy and make the item so that all the children could try it. They were required to work within a limited budget, with the rather sparse equipment in the school, so that the majority of the preparation could be done in the classroom area where supervision by the teacher was possible. Because the children were required to work fairly independently and in a very practical activity, it was noticeable that they persisted, discussed and solved mathematical problems with greater motivation than in other more overtly mathematical tasks.

The design of environments often requires practical measurement and the application and interpretation of scale. As with most fairly open-ended technological activities the level of mathematics can be varied to suit the mathematical development of the children involved. Some nursery children who were wallpapering some screens for a role-play area were given the problem of how to cut the right size of wallpaper. They decided to count the number of hands that would fit into the width of the screen and cut the paper accordingly. In the design and subsequent making of a model estate a class of year 3 and 4 children wanted the buildings to be of the same scale so they chose the side of a match box to be the standard for one storey. In the case of a class who designed and built their own pond and nature area the existing school grounds had to be first accurately surveyed and the measurements transferred to a scale plan.

Children should not only be involved in identifying, planning and making products for themselves, they should also be deciding what calculations and measurements are necessary and how they might be done. It is through such practical and applied activities that children realise that basic mathematical skills are essential, reach a better appreciation of when to use the four rules of number, and really appreciate different weights and measurements, making estimation much more accurate. The desire to achieve the end product motivates children to strive to understand and apply the mathematics required.

ART AND EVALUATING THE AESTHETIC DIMENSION

'The success of business, industrial and professional enterprises depends increasingly on the way in which products and services combine functional and aesthetic requirements to satisfy the needs of the international community. . . . While factors such as reliability, efficiency and sensitive pricing become the norm for all manufacturers, the aesthetic quality of design becomes very significant in influencing choice.'[27] Children need to ask three main questions when evaluating products: Does it meet the original need? Was the cost of production satisfactory? Is it aesthetically acceptable? They should also consider the relative importance of each and whether they are in conflict. An item might meet the original purpose and be cheap but is an eyesore. In another case an attractive and effective product might be financially too expensive, have taken an inordinate amount of time to produce or have highly undesirable environmental side-effects. All designers, including the children, need to strive for the most appropriate compromise between these three components.

Trying to evaluate the aesthetic element of a product is very difficult because personal differences in taste are so significant. However, children need some guidelines that enable them to identify elements that affect the aesthetic properties of a product which they can take into account when evaluating manufactured artifacts, environments and systems, and when designing their own. The visual elements of art such as line, shape, form, texture and colour, as well as some of the principles of design including balance, proportion, harmony and contrast, can be used to provide such a structure for the children's observation and design.

Therefore learning in art makes significant contributions to design and technology capability by assisting in developing children's critical skills and judgements about the aesthetic dimension. It is also fundamental in improving children's drawing skills which are essential in producing sketches, working drawings, plans and diagrams, and giving children increased competence in handling a wide range of material.[28]

Observation

An object cannot be evaluated without looking at it carefully and identifying in what way it is attractive and whether it is functional. Observation should involve all appropriate senses. A product usually needs to look attractive but may also need to feel comfortable or pleasant. Food dishes should smell appetising and taste good, music should sound stimulating and a garden for the visually impaired should have many different interesting smells and a variety of shapes and textures to touch. In order to develop the children's observational skills and descriptive language they need many opportunities to look at and discuss existing products and should be encouraged to notice as much detail as possible and not be limited by what they expect to find. There will be a tendency to concentrate on artifacts but children should also examine systems and environments.

Open questioning and oral discussion are essential to develop the quality of children's observations. Whether the child is looking at an arrangement of flowers, piece of furniture, or the way a dance group moves together, the teacher's questions about colour, texture, pattern, etc. will increase the children's perception and help to develop their aesthetic awareness. It can also help if the children are asked to look for differences and similarities between more than one item of the same type, e.g. two teddy bears, two shawls or two flower beds.

Asking the children to record what they see in drawings, paintings, clay or another media will assist further. Such illustrations may be general views of the products or could concentrate on detail such as the line, shape or texture with annotations. The more children are encouraged to look at, touch and talk about what they draw the more detailed will be the observations and quality of the work. The teacher should concentrate on using questions to help the children to notice more and not succumb to the temptation of interfering too much with the actual recording.

Alongside comprehensive observations of products it is helpful to provide focused activities which enable the children to explore individual elements such as line, shape, form, texture and colour which will provide the language that enables them to communicate about the attractiveness or otherwise of a product, as well as helping them to learn to create the effect they wish to invoke when making their own products.

Line

Line is the simplest and most direct way of expressing ideas on paper. Initially children might experiment with different line-making tools such as chalk, crayons, pencils, brushes and fingers or try to find as many types of lines that can be produced with one tool: dotty, wiggly, thick, straight, curly, jagged, etc. Once they have some idea of the meaning of the term they can look for evidence of lines around them: varying thickness and disjointed curving lines on the grain in a wooden cupboard, upright thick sharp lines on the radiator and soft, indefinite lines of the stitching on a pullover. They might combine these experiences to choose an appropriate tool to record the lines they have observed.

The lines do not have to be made on paper: they might be made in two or three dimensions as in the girders of a bridge, iron bars of a gate or lines linking pylons. Other less obvious lines can be identified in many products such as the switches in rows on a hi-fi and sequins or buttons on a dress. The children could record these observations in a two- or three-dimensional way by using string, thread, straws, wood, wire, etc., and comment on how altering them in some way might improve the product or make it less attractive.

Shape and form

In art the term shape is used to denote a two-dimensional representation, while solid three-dimensional things are said to have form. When children first look at shapes or objects they tend not to see that there is a shape around them. Even a circle drawn on a paper makes a second shape. Children need to have their attention drawn to these relationships, as in a design it is important that there is a balance between the main features and background. The children can be helped to concentrate on this space between objects by using a view-finder (a rectangle of thick card with a small rectangle cut from the centre) to observe and then draw a small portion of a group of artifacts, such as a teapot, cups and dish, or they could be asked to draw only the space around the artifacts.

The children might also look around their environment to find different shapes and forms such as on windows, brick walls, paving stones, lampshades, drapery and furnishing; decide which are purely decorative and which have structural function; and suggest improvements. When they look at both natural and man-made forms such as pebbles, shells, flowers, ornaments, cutlery, furniture and toys they should be encouraged to examine them from all sides so that they consider several views when making their own products.

Some stylised shapes are deliberately made to look very simple and show the main features only. Logos, for example, are designed to represent an idea which will be readily identified and related, for instance, to a business, football team, school or nation. Street signs are intended to be understood easily with minimal words. Children could look at signs and symbols used in their environment, heraldry and religions and in industry and business to see how shapes are used to communicate a meaning without words. They can then invent a school sign, their own family 'coat of arms', club badge or T-shirt design.

Texture

Texture is particularly related to touch. When observing a new object it is often useful to put it into a box with a hole in the side or into a feely-bag for the children to feel before they see it, so that they are encouraged to concentrate on discussing the texture and shape. They might also take rubbings or make impressions by pressing plasticine onto different objects, including fabrics, street furniture and walls, to help them notice the variety of textures that they probably had not realised were there. Although different materials have their own appearance and feel (wood has a distinctive texture and tends to feel warm whereas metal usually feels hard and cold), the texture can be altered during manufacture, such as by sandpapering wood and by deliberately rippling the surface of concrete before it hardens. The children could look for such examples and comment on whether they feel this has improved the look or function of the product.

Texture is not only felt but is also seen. A rough surface casts very small shadows so the texture is seen by the contrasts of light and indeed can be accentuated by judicial placing of a bright light. By using the idea of contrasts of light and dark, flat surfaces can be drawn or printed to seem textured as in some wallpapers. Different textures found in the environment can be recorded in a wide range of media such as pencil, pastels, ink and chalks, and the most attractive used as a basis for make printing blocks by glueing string on the thick card to print wallpaper or book covers.

Colour

As children develop an appreciation of the variety of tones within any one colour in their environment, they can be encouraged to notice how colours vary when in light or shade, and how texture influences the same colour. For example a rough material and a shiny one dyed in the same colour will have a different look. The children also need to explore the effect of the juxtaposition of colours. These observations are important early experiences in the design of many things including clothes, the relationship of paintwork and upholstery on cars, and the way colours are used together on posters and notices. In some products it is important that two colours are harmonious whereas in other situations a contrast is required. Most people want the colour schemes of their rooms to be compatible, whereas it is important that safety bands fixed to clothes stand out so that they are seen at night, and combinations of colours in road signs should be carefully chosen for maximum visibility. The children can investigate which two colours stand out well by attaching the same shape in different colours to identical backgrounds and then holding the signs at a distance to see which is most visible. They should find that those colours on the opposite sides of a colour wheel of the spectrum have greatest contrast. The children are also likely to discover that

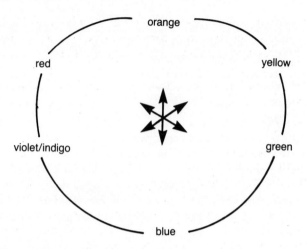

Figure 3.1 Contrasting colours of the spectrum

tones of the same colour and colours adjacent to each other in the spectrum are most harmonious.[29] (See Figure 3.1.)

Colour is very important in terms of establishing mood. Whether by innate psychological response or by cultural tradition we often respond to colours in different ways. The fashion industry and advertising often build on this, showing images of red cars to indicate speed and excitement and using white and blue to denote purity and cleanliness. It is important to remember that these responses are not exactly the same in all cultures. Red, for example, is often used to communicate messages to do with aggression and danger whereas it is also considered to be especially lucky in both the Hindu religion and the Chinese culture.

The children could investigate whether different colours make people feel differently. Groups of children could be asked to paint a happy picture or backcloth denoting sadness for a dance or drama, using only colour, shape or pattern. The task could involve a survey to find out what colours make people sad or happy and might lead on to evaluating colour schemes used in the school and home. When considering colour schemes in the design of specific environments the children could examine a collection of wallpapers, choose one and say why it appealed to them in terms of pattern, texture and colour with a view to using it to decorate a role-play area. They could even print their own wallpaper using large junk items, rollers and polystyrene for printing.

Colour is also significant in the design of food dishes. This is partly because the diner wishes to have an attractively presented dish but also because we are wary of unusual colours in food. The children could investigate how colour affects their food likes and dislikes by making up a packet of mashed potato and colouring it with different food colouring; then asking others to taste it, seen and unseen, and make comments. They will probably discover that certain colours are unpopular even although the taste is the same. A similar activity involves first choosing a favourite sweet from a packet of sweets and then trying to accurately identify the colour blindfolded to see whether preference was really related to the colour or the taste. The children could also write to a sweet manufacturer asking about colour choice in order to use their research to make and sell their own sweets.

Pattern

Line, shape and colour can all contribute to making a pattern which has a repeating element in it. As children explore products in their environment to discover patterns they should try to identify which have special functions and which are for embellishment. Man-made patterns of bricks in walls and tiles of roofs are primarily functional, related to creating structural strength, but can be designed to be attractive as well. Other patterns are purely decorative such as in printed fabric, repeated carvings in furniture, geometric patterns in mosques and recurring symbols on linoleum. Children need to have opportunities to make their own by painting or printing, repeating shapes using junk and other items, initially on paper and later on fabric and other media. Once they understand how to produce pattern and its effect they should be able to apply their knowledge when creating technological products.

Balance

The children can investigate ways of creating balance and distinguish when it is used for functional or aesthetic purposes. Functionally it is useful to have balanced structures which fit the basically symmetrical human body and are easy to construct. Bilateral symmetry, where one-half of an object mirrors the other half, is a very common way of providing balance and is to be found in many manufactured products including some furniture and clothes. Other designs use radial symmetry where the balance radiates outwards from a central point as in a wheel, circular table or flower bed.

However some people find such simple balance rather boring, so in order to provide a more unusual or stimulating design the product could be asymmetrical but still balanced. In this case the left and right side of a design may not be the same but are balanced visually or in terms of the forces acting on the object. As part of work on stability and the centre of gravity the children may design toys or mobiles that balance on the edge of a table or on a wire. A parrot shape can be curved in such a way that its shape is evenly balanced around its perch but looks unsymmetrical.[30] Not all products need to be balanced for functional reasons, such as a book-cover design, poster or garden, and in these cases a deliberately discordant or unbalanced effect may be attempted in order to excite, shock or stimulate.

Harmony and contrast

In the same way, combinations of lines, forms, patterns, etc. can be harmonious or contrasting. A contrast may be created using thick black lines with fine ones; linking geometric shapes with imprecise shapes; juxtaposing rough textures with smooth; and so on. Some contrasts work well together whereas others give a sensation of discordance. The designer should deliberately choose combinations to create the desired effect.

Proportion

In order to provide comfort some proportions are very significant. As most products are intended for human use the dimensions and movement of the human body are very important. People need to be able to fit through a door, look out of windows, reach the pedals on a bicycle, hold a handle of a suitcase, step up a flight of stairs, and both listen and speak into a telephone. Producing the ideal design is difficult because people vary. The children might investigate the environment of the school and home to identify which things have been made with adults in mind and which for children. They might consider whether chairs, desks, beds, cupboards, display boards, sinks, taps, door knobs and so on are too high, low, large, small, wide or narrow for their comfortable use. From this experience they could design an environment to cater for children, people in wheelchairs or parents with pushchairs.

All objects are placed in a context, and proportions of adjacent items can seem right or wrong. Some of these impressions will have been built up through repeated experience of the environment and of natural forms creating an expected familiarity. Other combinations are unacceptable because they are impractical. Toy animals with unusually long limbs or large bodies jar the senses, a nursery chair next to a full-size table is not only impractical but it also looks out of place, as might a small building surrounded by high-rise flats. The designer has to decide whether such combinations enhance or stimulate the look of the product or detract from its appearance. There is no correct answer in many design decisions but by being aware of the significance of proportion the children are better able to analyse their feelings and impressions and use this knowledge to deliberately create the impression they want.

The children need to appreciate that aesthetics is not concerned exclusively with appearance, texture, smell, sound and taste but the way the product works also influences the user. The magnetic 'pull' of the fridge door, the ease of movement of a volume-control button and the accessibility of the radio in a car all influence the feeling towards the product. A good designer has a broad approach to the design of a product, including considering the effect of the product on all the senses of the clients and their emotional responses, as well as attempting to match the product as closely as possible to their physical and functional requirements.

NOTES AND REFERENCES

1 Williams, P. (1990) *Teaching Craft, Design and Technology: Five to Thirteen* London: Routledge.
2 Liddament, T. (1992) 'Teaching design through social and historical contexts' *Design and Technology Teaching* Vol. 24, No. 2, pp. 18–21.
3 National Curriculum Council (1990) *Curriculum Guidance 7: Environmental Education* York: NCC.
4 National Curriculum Council (1990) *Curriculum Guidance 4: Education for Economic and Industrial Understanding* York: NCC.
5 National Curriculum Council (1990) *Curriculum Guidance 8: Education for Citizenship* York: NCC.
6 Bell, S. (1990) 'Radio activity' *The Big Paper* No. 9, pp. 10–11.
7 Hissey, J. (1986) *Old Bear* London: Beaver Books.
8 Hughes, T. (1968) *The Iron Man* London: Faber & Faber.
9 Armitage, D. and R. (1977) *The Lighthouse Keeper's Lunch* Harmondsworth: Penguin.
10 Campbell, R. (1982) *Dear Zoo* Harmondsworth: Penguin.
11 Proysen, A. (1968) *Mrs Pepperpot to the Rescue* London: Penguin.
12 Williams, T. (ed.) (1991) *Stories as Starting Points for Design and Technology* London: Design Council.
13 Hislam, J. and Jarvis, T. (1992) 'Bridging the discipline gap: Linking story themes with science and technology' *Education 3–13* Vol. 20, No. 2, pp. 38–44.
14 Fisher, R. (ed.) (1987) *Problem Solving in Primary Schools* London: Basil Blackwell.
15 Smith, R. (1990) 'Stories from science' *Primary Science Review* No. 13, pp. 4–5.
16 Johnson, P. (1990) *A Book of One's Own: Developing Literacy Through Making Books* Sevenoaks, Kent: Hodder & Stoughton is a comprehensive guide to book making for primary children.
17 Hiner, M. (1985) *Paper Engineering for Pop-up Books and Cards* Stradbrooke, Diss, Norfolk: Tarquin provides 10 basic mechanisms which can be used as starting points for further inventiveness. Gibson, R. and Somerville, L. (1990) *How to Make Pop-ups* London: Usborne provides examples of a range of different projects.
18 Cousins, L. (1990) *Maisy Goes Swimming* London: Walker.
19 Byron, K. (1989) Tunnelling for knowledge *Questions* Vol. 2, Issue 2, pp. 6–8.
20 Kitson, N. (1986) Drama: A context for cross-curricular learning *Primary Education Review* No. 26, pp. 8–11.
21 Littledyke, M. (1989) Employment versus pollution *Questions* Vol. 2, Issue 3, pp. 8–9.
22 Jarvis, T. (1991) 'Primary technology: The value of evaluating and improving environments' *Education 3–13* Vol. 19, No. 2, pp. 23–29.
23 Potter, M. (1990) 'Problems for Severus' *Questions* Vol. 2, Issue 6. pp. 14–15.
24 Potter, M. (1990) 'Smashing the new machines' *Questions* Vol. 3, Issue 3, pp. 46–7.
25 Littledyke, M. and Baum, C. (1986) 'Structures for drama: The market place' *2D* Vol. 5, No. 2, pp. 72–7.
26 Pinel, A. (1990) *Signs of Design: Mathematics* London: Design Council.
27 DES (1991) *Art for Ages 5 to 14* London: HMSO, para. 3.38.
28 Ibid., para. 3.41.
29 Breckon, A. and Prest, D. (1983) *Introducing Craft Design and Technology* London: Thames/Hutchinson has very useful sections on colour and other visual elements.
30 Johnsey, R. (1990) *Design and Technology through Problem Solving* London: Simon & Schuster has a very useful section on the design of balancing models.

FURTHER READING

Keightley, M. (1984) *Investigating Art: a Practical Guide for Young People* London: Bell & Hyman.

Milloy, I. (1990) 'Craft design technology in the primary school: Let's keep it primary' in Bentley, M., Campbell, J., Lewis, A. and Sullivan, M. *Primary Design and Technology in Practice* London: Longman.

Rowswell, G. (1983) *Teaching Art in Primary Schools* London: Unwin Hyman.

Ruff, P. and Noon, S. (1991) *Signs of Design: English* London: Design Council.

Chapter 4

Evaluating and improving environments

As with all design and technology products children need to learn to observe and evaluate existing environments, identify and suggest ways for improvement, and then plan and carry out their ideas. Additionally they are expected to evaluate the success of their work. This chapter considers an approach to help children observe and analyse their surroundings, followed by examples of how children might plan and carry out improvements, initially in very familiar environments such as their homes and classrooms and then in increasingly unfamiliar localities.

The study of environments in technology has particularly close links with geography as both involve the observation and analysis of localities and the examination of the influence of humans on their environment. Many geographical and technological projects can be interrelated in such a way that the children's appreciation and understanding of both subjects can be enhanced. Additionally the ability to draw and interpret plans and maps is an essential skill for the two disciplines in order to communicate ideas about places. As a first step to understanding cartography children usually make simple plans of familiar locations such as their home and school. This can be extended into a technological activity by asking the children to use these plans to discuss what features they like or dislike and suggest improvements or to design an ideal room. Once the children have a basic appreciation of plans, skills of reading and producing a wide variety of maps, understanding and applying cartographic conventions such as keys, coordinates, scales and directional information can be developed and improved as part of many technological environmental projects. However it should be remembered that plans are not a substitute for first-hand experience of different environments but are primarily a means of enriching communication of observations and suggestions for improvements.

OBSERVING AND ANALYSING FAMILIAR ENVIRONMENTS

One approach to evaluate an environment is to focus on observing and describing specific aspects and then to analyse these in terms of their function and aesthetic purpose, in order to suggest changes.

Focused observations

Getting children to look carefully at a very familiar place, such as their classroom, school or nearby street, can be difficult because they feel they already know it well, but this can be achieved by asking them to concentrate on an unusual view or limited focus. The children could be asked to search for things they have not seen before; examine features on the skyline or above their heads only, and find detail of items at ground level; look at things from an unusual angle such as from above or below; concentrate on the sense of touch, sound or smell; or focus on specific elements such as colour, shape and pattern.

The children also need to be questioned and encouraged to talk about what they notice, to compare features and to try to find ways of describing what they see. Recording their observations will further motivate them to notice detail. The children might produce plans, maps, descriptive pieces of writing, poems, annotated drawings, sketches, three-dimensional models, rubbings of different textures, or tape recordings of actual sounds or musical representations of the typical noises of the locality.

Identifying and evaluating functional and aesthetic factors

In order to evaluate the environment either the children could concentrate on the individual elements that they have picked out or they could discuss the area as a whole. They need to identify the purpose of the feature or place, determine whether it is primarily functional or aesthetic and decide how successful it is.

The children may have been asked to find examples of different patterns in the school. They should find that some are mainly practical, such as the interlocking bricks designed to give strength and the lines of coat pegs for optimising storage space, whereas the pattern on curtains is entirely decorative. In the same way some shapes are principally functional, such as rectangular windows, which are quicker and simpler for a builder to use, and the strong triangular shape of roof frames, whereas shapes within the patterns on wallpaper and vinyl are ornamental. Textures can also be functional, as in textured paving at the edge of a pedestrian crossing, smooth easily-washed floors in a wet area, and soft carpet to reduce sound and provide comfortable seating. The children might then suggest ways of making improvements. Children are not going to be able to make a complete analysis of all elements in an environment, but by examining some in this way they will become more aware of the necessity to consider both purpose and appearance of designs.

When identifying the functional and aesthetic aspects of an environment as a whole the children again need to decide what are the purposes of the area and how well this is achieved. Their local street is probably intended to give efficient access to people in vehicles, on bicycles and on foot. Some of the people will be travelling through the area, whereas others will want to get to shops and other buildings. In order to decide how successful the design is for each of the identified purposes the children might ask questions about each category of user such as: Are the pavements wide enough for people with large families and push chairs? Are there safe places to cross the road? Are these in suitable places? Are there places for people to wait at bus stops without taking up the whole of the pavement? Can they wait in shelters? Do stopped buses cause traffic jams? The children also need to consider the positive and negative aspects of the visual and sensory appeal of the area and may refer to the standard of upkeep of the buildings, cleanliness, amount of litter, noise and smells. Having focused on aesthetics and function the children are likely to be able to suggest improvements in both.

Awareness of change

Young children tend to assume that solid buildings do not change. The children could be encouraged to remember minor changes in their home or school, perhaps an extension, the reorganisation of the classroom environment with a change of teacher, new furniture replacing old, and minor building improvements to the interior or grounds. They could also make a note of changes in their street which may include changes of a shop front, trees planted or removed, demolition work and building sites under development. Evidence of past change can be found by examining old maps and photographs, and clues (such as bricked-up windows) may be found on existing buildings. Once the children appreciate that there is constant change in most places they might consider who causes the change and why. Individuals can help to control the level of cleanliness and litter; a shop owner may make modifications to attract more people into the shop; and the council makes alterations to increase safety, improve services and make better use of technology. It is important to realise who is responsible for change so that the children know whether they can make changes on their own or whether they need to persuade others to act for them.

Technological projects should be set in a variety of contexts that the children can relate to in the home, school, recreation, community, business and industry. On the whole the youngest children will evaluate and design everyday environments based on their homes and school, and when they become more mature and experienced they will study localities which are more complex and unfamiliar. However this is not a hard and fast rule as the early years child can successfully examine small local businesses, and challenging opportunities for the older Key Stage 2 pupil can arise from study of the home or school.

HOMES

Initially the children can look at the design of their own home, bedroom, kitchen and living room. They might draw their own room and discuss what features they like or dislike in order to design the ideal bedroom. Older children could be given furniture catalogues and wallpaper books and be asked to work to a specified budget. The design of different rooms can also be considered so that the children could plan and make a role-play kitchen, sitting area or bedroom for the classroom area.

Children can build on their developing knowledge and awareness of different interior environments to produce designs for specific purposes. Very young children could produce pictures or models of a home for a mythical animal, person in a familiar story or a caravan holiday home for a doll. A witch, a creature from outer space and a giant would all need very different homes. The children should be asked to think about the features and needs of the individuals in terms of size, strength, sleeping, eating facilities, etc. and try to accommodate each requirement. Older children could design a room or building for physically disabled people, a child or someone old, which could be researched by visiting an old people's home or finding out what equipment is already available for the disabled. Another project could involve the children finding out about 'environmentally friendly' products, modern methods of heating and insulating homes in order to design a home for the future.

The interior and exterior designs of buildings from different historical times can also be studied. Ideally this is best developed from a visit which may be primarily intended to establish historical skills and knowledge, but technological activities can also be included. The children can be helped to appreciate that the design and architecture of the interior and exterior of buildings from different historical times will have been constrained by what materials, equipment, tools and energy sources were available, in the same way as their designs of present-day rooms are. For example the difference in heating method influences size of rooms and windows, need for a fireplace, location of furniture, floor and wall covering, etc. Similarly garden design and farm planning were related to available animals, plant species, agricultural equipment and social structure. For instance before the industrial revolution fields were small and enclosed by hedges, whereas modern farms maximise the use of large machines by having very large fenced areas. Such an approach enables historical artifacts and environments to be seen as not to have just happened by chance but to be a logical response to level of technological achievement and the organisation of society which occurs in exactly the same way today.

The design of homes reflecting different cultures are also responses to available materials and equipment as well as to the lifestyle and needs of the families concerned. Children in the class with personal experience of different cultures and countries should be able to give valuable information to their teacher and peers. Hindu children, for instance, are likely to have small shrines in their homes, and kitchen environments are likely to differ as a response to the different cooking methods. Obviously these discussions need to be approached in a sensitive way emphasising that a design is successful, not because it is familiar, but because it meets the special needs of that family. The design of different homes in the world can be studied in a similar way. For example homes in Afghanistan and Spain are different from those in Britain, but are well designed for the hotter climate by having flat roofs where people can sleep safely; a few small windows to keep out the sun's rays; and open-air courtyards which act as outdoor rooms where the air becomes cool at night and stays cool in the daytime. The materials used should not be automatically regarded as less advanced or bizarre because they are not the same as those the children are used to but, like the clay adobe houses in the USA and Arabia that absorb heat and release it during the cold nights, they usually have ideal properties for their particular situations.[1]

THE SCHOOL

Redesigning and improving classrooms

Teachers spend considerable effort in arranging their classroom to maximise the children's learning. This process can be effectively achieved with the teacher and children working together. One student-teacher who had just moved to an inner-city school did not feel the classroom was set out in a way that she could cope with. She particularly wanted a large enough area where the children could sit on the floor so that she could talk to them all together. The children were also located on tables arranged in very big groups, whereas she preferred smaller, more flexible groups. The student explained this to the children and encouraged them to share the features they wanted. Following the discussion, pairs of children prepared alternative plans. These were evaluated together and the best features of several were combined and the new arrangement installed. The student reported that not only was the new arrangement very satisfactory but also the exercise had unexpectedly helped to develop a mutually supportive atmosphere in the class. Other successful projects include shared decisions on how to incorporate a woodwork area, book corner and wet area suitable for science and mathematics experiments. Reorganisations of the classroom do not have to be on such a major scale. Smaller ways of improving the classroom may be suggested by the children, such as adding pot plants, providing exciting displays, printing fabrics for drapes, making cushions for the reading area, or covering boxes to keep equipment and materials tidy.

Some of the changes may be intentionally temporary as in setting up a role-play area or converting the classroom into a scene related to the current project, so that the children can have several opportunities to design and actually create new environments building on past experiences. These are particularly important as it is often difficult to enable children to go beyond planning and modelling new environments in many other situations. Children throughout the primary school can participate in the planning and subsequent work as demonstrated by a group of four-year-old children who planned and set up a new home corner, including wallpapering dividing screens in their nursery. In another school Year 2 children set up a clinic after a visit to their nearby health centre which incorporated a bed, home-made instruments for the nurse and reception area with magazines and comics. As part of a history project 10-year-old children designed and created a scene at a Viking harbour with a longboat as the centre piece. Other undertakings have included creating different types of shops, an underwater setting and an imaginary scene based on the giant's castle from 'Jack in the Beanstalk'.

Animals in the classroom

Pets that are kept in the classroom can also provide the challenge to provide suitable environments for specific purposes. If fish are kept the children can be asked to find out about their requirements and plan their habitat. The types of cage for a rabbit, gerbil or hamster can be decided upon by the children and they should take charge of ensuring that this home is kept clean. In one school a group of five year olds decided that their gerbils needed a bigger home and temporarily converted the sand tray. They provided sleeping and living areas and used wire netting to stop the gerbils escaping. In similar ways the children can help set up suitable temporary homes for tadpoles, worms and caterpillars.

The school interior

In order to enable children to feel confident that their suggestions to improve environments around the school will be valued it may be necessary to suggest specific tasks. They could be asked to make the school more welcoming to visitors. Suggestions made by children have included improving the entrance hall with pictures and plants, providing notices in different languages, putting up photographs of the secretary and teachers and indicating where to find them, and providing a play-pen and toys for very young children who visit with their parents.

Improving the school grounds

In the past many schools have made improvements to their school grounds to provide a rich environment for science projects and to provide stimulating surroundings for the children. Unfortunately in the past the children have infrequently been involved with the decision-making process and

Table 4.1 Percentage of schools, for each authority, that considerably involved children in the planning and installation of a nature area

	Birmingham	Leicestershire	Northamptonshire
Planning	13	31	35
Installation	26	43	33

Table 4.2 Curricular subjects supported by the nature areas in Birmingham, Leicestershire and Northamptonshire schools in 1991

	Birmingham	Leicestershire	Northamptonshire
Science	62	63	74
Art/craft	48	43	55
English	39	42	44
Mathematics	24	30	37
Geography	23	18	29
Technology	21	16	28
Other	17	8	7
No. schools replied	63	63	74

In Leicestershire the planning authority is aware of about 100 school nature areas, so the reply rate probably represents a fair picture of nature areas in Leicestershire. A similar proportion was received from the schools in Birmingham and Northamptonshire.

implementation. However this is changing, as demonstrated for example by schools in Birmingham, Leicestershire and Northamptonshire. Questionnaires were sent to all these schools in 1991 asking for information regarding the planning process, installation and use of nature areas. In Leicestershire and Northamptonshire about a third of schools had considerably involved their children in the planning and actual installation, whereas Birmingham schools had involved their children less. (See Table 4.1.) However there has been a steady increase in the children's involvement over the past few years. Five years ago most schools in Birmingham did not include their children at all in the planning process, whereas now many do.

However even with this increased involvement of children in setting up a nature area few schools see it as being used to develop design and technology. Of all the subjects thought to be supported by nature areas technology was the least. (See Table 4.2.)

In order to start to improve the school grounds it is important to assess the potential of the site for short- and long-term projects. The children can be involved in this assessment to evaluate existing arrangements and potential areas for change and in trying to raise funding. Even where there are limited grounds or where the land is taken up by playgrounds and sports pitches there may be opportunities to develop the boundaries, provide hanging baskets or potted plants, or add murals and sculptures to the walls. Plans of existing provision can be drawn up with relevant details that may influence what new provision can be made such as the microclimate (Is the area a natural sun trap, likely to be wind swept or subject to frosts?) and whether there is access to water, drainage or electricity. Many local education authorities have a register of environmental schemes and give advice at early stages. Voluntary bodies will often also give advice and help with funding is occasionally available.[2] With most projects the children should be able to be involved in the construction and installation process as well as with planning.

It is generally better to have a series of small projects than one major enterprise, as they are easier to finance, will cater for several groups of children over the years, and are likely to be realistic about staff resources and commitment. One school, for example, sustained a series of fairly small projects which lasted over 12 years and are still ongoing. These started with a greenhouse financed by the sale of waste paper and included a chequer-board garden designed as a mathematical exercise, establishment of a pond, planting of trees and shrubs, butterfly house and provision of seating for children and parents. The gradual process meant that the staff were not overwhelmed or discouraged.[3]

Some schools have made their playgrounds more stimulating by painting lines on the ground for games from different cultures, like hopscotch, and making interesting shapes for imaginative play; producing additional equipment for the school grounds, such as benches, shelters and climbing apparatus; and providing facilities to attract birds, such as nesting boxes, bird tables and planting seed-bearing shrubs and plants. Other successful projects have included making small gardens for flowers, vegetables or plants that attract butterflies and other insects; establishing wild-flower meadows;[4] planting trees on school boundaries; establishing small woodland areas; and digging and stocking ponds.[5] A pond is a very ambitious project and is an excellent resource for science, but many Birmingham, Leicestershire and Northamptonshire schools reported that they were less likely to be successful in the long term if cheap methods of lining are used. It was generally felt that, in order to limit problems of maintenance, ponds need to be appropriately built with high-quality materials.

Maintaining and cleaning up school grounds

A natural resource modifies and changes continuously so on-going monitoring and maintenance is essential and can be carried out as technological and science projects. The children might identify the characteristics of weeds, record their spread and take action to control them. Understanding the reasons and methods of maintenance are ways of showing children how ecological concepts are applied practically. Problems will also arise which they can help to solve, such as the pond becoming clogged with leaves. In this case the children will need to find a way of clearing the pond and then to identify where the leaves are coming from in order to stop them blowing into the pond. Other technological projects could include designing and making traps to study minibeasts that will not harm the creatures, and setting up a nature trail with an associated information pack.

Part of the process of caring for the grounds also involves keeping it clear of litter. As the science curriculum requires young children to investigate how far everyday waste products decay naturally and to use this knowledge to improve the appearance of their local environment, this can be combined with a technology project. After having carried out a survey of litter in the school grounds to assess the extent of the problem, the children can attempt to explain their findings so that they can suggest different ways to reduce it. These might include making litter-collecting devices, creating eye-catching litter bins that are designed to stop litter blowing out but are also easy to empty, or producing posters encouraging people not to throw litter. The success of these solutions should be reviewed, improved and evaluated, perhaps by comparing litter surveys before and after intervention.

RECREATION

The park and playground

To help children identify some of the design factors in a park, they could be given a simplified plan of a hypothetical location without any facilities marked, such as a large grassed area crossed by a path bounded by a stream and wooded area. The children can then be asked to add a flower garden, swings, football field, roundabout, slide, sandpit, seats, model-boat pool, litter bins and climbing frame, giving reasons for their arrangements. By asking who the main users of the area will be and how their needs might differ, and which of the fixed and movable items go together and which should be separate, the children should be able to make logical decisions. It is likely that children, their carers, young footballers, and old people will use the area most. The children will want easy access to the play equipment and the carers need places to sit where they can supervise easily, whereas other adults may want peace and quiet away from the children. Some benches therefore need to be near and facing the play areas with litter bins close by, and other benches in more scenic positions. The play equipment needs to be located carefully for reasons of safety and it probably would not be wise to have the sandpit too close to the pool. Discussions along these lines should enable the children to see that the various users of the park will want different facilities and that careful location is necessary to allow each group to enjoy the area without interference, and in some cases an ideal solution may be impossible.

Following an activity to heighten the children's awareness of features to observe, a visit to a park could be arranged to evaluate how similar design questions have been tackled. The children could comment on what aspects they find attractive and what they liked or did not like, with a view to suggesting improvements or designing a new park. The project might be extended for older children by asking them to suggest how to change an existing park so that it can be enjoyed by the visually impaired and physically disabled children.

Designing indoor recreational areas

Key Stage 2 children need to increasingly think about the needs of others and can also consider general issues of health and safety. The children might plan a holiday playcentre to cater for both boys and girls of different ages which includes finding out what activities different children prefer; planning to set out equipment safely; providing sufficient washing and toilet facilities; and ensuring young children can be adequately supervised. Designing a parents' room could entail asking parents to visit and discuss their requirements and interests which should help the children to become more aware of the needs of adults. Other exercises can include evaluating an existing museum after a visit, by considering whether the displays appeal to a wide audience, how effectively the information is given, whether disabled people are being accommodated, and whether catering and retail facilities are appropriate. The children might then set up their own museum area using artifacts and their own work that relates to a historical project, in an attractive and exciting way, with information given in different languages, on tape and by using videos.

When studying a relatively unfamiliar environment it can help to start off by looking at an imaginary example with many faults so that the children see some of the factors involved in the design. For example the children could be given the plan of a disco with many fire risks, such as a narrow staircase as the main outlet, limited supplementary exits, coat pegs fixed near a cooker, heaters located near doors, electric points for the amplifiers at a distance from the stage so that wires are left trailing across the floor, and so on. The children could then be asked to list the fire risks in order to decide whether they would give the new disco a fire safety certificate. Having thought about the facilities required and having had the issue of safety highlighted, the children could finally design their own disco.

THE COMMUNITY

Investigating the neighbourhood around the school

As part of a project to heighten the children's awareness of safety, the area around the school could be investigated to discover features that promote safety, such as road signs and pedestrian crossings. Dangerous areas could be identified and ways of reducing the risks explored. Proposals can be illustrated on plans, annotated drawings and models, and might include adding road humps and increasing the culs-de-sac to reduce the number of cars travelling through the neighbourhood at speed. Key Stage 2 children need to appreciate that the range of criteria which must be used to make judgements about what is worth doing may be conflicting, and they need to find satisfactory compromises. In this case car users will not be pleased to lose all their short cuts to avoid traffic jams, so it may be necessary to include a fast route with pedestrian footbridges.

The appearance of their locality can also be evaluated by the children marking a street map with comments on places they feel are either attractive or unattractive. Suggestions could then be made on how the area can be improved, perhaps by removing graffiti and litter, repainting buildings, painting murals, providing bins and planting flower beds. As part of the overall ambience the children could also survey noise pollution by listing noises heard, and their level and frequency (loud and continuous, not so loud but continuous, loud but intermittent, not so loud but intermittent, no apparent sound), or use a decibel meter borrowed from a local secondary school. Ways to cut out undesirable noises could then be investigated, for instance by gauging the effects of putting a clock in a box surrounded by newspaper, polystyrene, fabric pieces, shredded paper, bubble plastic or straw in order to assess which material is the most effective sound insulator. From this knowledge the children can design comfortable earmuffs for people working in the noisy area. They might also consider the possibility of insulating buildings or writing to the organisation making the noise.[6] However they may have to accept that change would be very difficult to achieve, either because the costs are too high or because there is insufficient public pressure to force a change.

Housing estates

As children get older they are able to consider problems or environments that are increasingly complex and not so familiar. As a combined geography and technology project a group of children aged 7–9 years were asked to act as architects in order to design a housing estate for 1,000 people. They were given a completion date on which they were to give a presentation and show their plans to the 'Town Clerk', i.e. headteacher, who chose one to be made in model form. The children were first asked to suggest advantages and disadvantages of existing house types (flats, terraced, semi-detached and detached houses) and asked which they would prefer living in. They also looked at plans of each type to compare the amount of space taken up by the house and garden area in order to decide which to have in their estates. The children also collected information about homes in their area from estate agents and wrote a specification for their own house to include comments about facilities in the area. As a side issue, therefore, the project enabled the class to discuss aspects of advertising.

The children were taken around their area to study existing facilities, such as shops, parks, pedestrian crossings, schools and factories and encouraged to list features that a community would need and which should be located together. They suggested that the mosque, health centre and park should be located together but the factories should be isolated; that the area as a whole should be made attractive with flower beds and trees; and that there should be pedestrian precincts to stop cars using the area as a rush-hour route. The initial quality of the plans varied enormously, but after group evaluations many were readjusted and the final standard was very high. The children then produced a large model of the chosen plan, which led into considering how to make the different buildings to scale and ways of constructing street lights, fences, etc. This technological approach enabled the children to apply their geographical knowledge and skills in a realistic way and had the effect of helping them to appreciate the relationship between three-dimensional locations and their two-dimensional plan equivalents, and the usefulness of such representations.

Religious centres

Visits to different religious centres as part of RE enables the buildings to be compared and the needs of people of very different cultures considered. The children can be guided to see that these environments are a response to the different religious beliefs and approaches. The mosque only has geometric patterns as Muslims believe that they should not represent the form of Allah, and as man was made in the image of god, statues and pictures of people would be unacceptable. The absence of chairs or seats in the gurdwara reflects the way Sikhs worship. The brightly decorated Hindu temple with images of god is again very different but is still a direct reflection of the needs of the worshipping people.

BUSINESS AND INDUSTRY

Shops and shopping precincts

Business environments are not only usually complex, but commercial factors are also significant. A shop is both a system and an environment to sell the maximum number of items at a profit. A visit to a local supermarket would enable the children to identify the different features, including cash desk, shelving, freezer, aisles, price labels, customer doors and delivery doors. Once these have been listed the children can be encouraged to explain their location and design. Questioning will also help the children to identify design principles that are intended to optimise sales, such as the need to present goods attractively; making all products reachable and easily transported by shoppers and staff; and placing luxury goods at eye level or at the checkout to tempt the customer, whereas staple commodities can be placed at less accessible locations as customers will search them out. Once the children have identified what they think makes an effective shop they can apply their ideas by suggesting improvements to an existing shop, setting up a role-play shop or designing a mobile shop. The work can be extended by comparing shops specialising in different goods such as food, furniture, clothes, shoes and musical instruments.

An existing shopping complex could be visited and evaluated with the intention of suggesting improvements or for designing a shopping precinct from scratch. The facilities, general ambience, way of making the area user-friendly, and safety can be reviewed and improvements suggested. A survey of the range of retail outlets, services such as banks, places for entertainment, cafés, public toilets, telephones and car parking can ascertain which are adequate or over-represented and whether they are appropriately located. Ideas for making the area attractive and user friendly could include providing attractive street paving, seats, hanging baskets, trees and shrubs, clear directional signs, good access and ease of movement for disabled people, seats for the elderly, mother-and-baby rooms, computer-controlled linked signs directing cars to multistorey car parks with spaces, large waste bins and easy access for emergency vehicles.

The farm

In similar ways a visit to a farm can initially lead to the children identifying the purpose of the main features and then trying to explain their location with the intention of producing their own designs. This could be at a very basic level intended to help children to understand that the farm buildings are located together for easy supervision, or aimed at a much more advanced understanding of how climatic and physical limitations combined with market requirements control both the type of enterprise and layout. The children could discuss the differences between different types of farm, e.g. a cereal farm and dairy farm, and those with knowledge of farming in other countries can contribute from their personal experience. The children could then use toy farm animals to set out a farm, design one on paper, or make a model.

Tourist industry and holiday resorts

A topic on holidays could lead to several technological activities that would enable children to investigate the tourist industry. They might design and set up a role-play travel agency. A plan of a seaside resort could be provided and the children asked to plan the location of different amenities, e.g. fishing, water skiing, volley ball, beach bar and disco. As some of these pursuits are incompatible the children should again appreciate that one provision affects other things and that it may never be possible to achieve a perfect solution. School visits to nature reserves, zoos, theme parks, concerts and theatres can also provide opportunities to evaluate other environments with regard to how well they cater for a wide range of customers including children, parents, teenagers, old people and visitors from overseas.

Work on evaluating environments increases the children's awareness of the quality of their surroundings and how they can act to enhance them. The process of improving the school and its grounds frequently leads to an increased sense of belonging, more responsible behaviour and a general reduction in litter, graffiti and vandalism.[7] In addition, as the children investigate environments in their neighbourhood, the increased contact with community activities, local business and industry may be able to help create greater mutual understanding, and assist the children to appreciate the constraints of the adult world, though also indicate that it is possible for every individual to have a beneficial influence.

NOTES AND REFERENCES

1   James, A. (1987) *Homes in Hot Places* Hove, East Sussex: Wayland; and James, A. (1987) *Homes in Cold Places* Hove, East Sussex: Wayland provide useful material on design aspects for children.
2   Anon. (1990) *Getting Help for Community Environmental Projects* Great Barr, Birmingham: Shell Better Britain Campaign. Learning Through Landscapes, Third Floor, Southside Offices, The Law Courts, Winchester SO23 9DL is a voluntary organisation that aims to help schools improve their grounds as an educational resource and for play and social development.
3   Nabney, G. (1991) 'Curriculum guidance seven and school grounds' *Environmental Education* Vol. 38, pp. 7–8.
4   The seeds can come from a commercial supplier of conservation mixes, or seeds can be collected if they are abundant. English Nature produces a booklet which lists the 62 species that are protected by the Wildlife and Countryside Act 1981.
5   Anon. (no date) *Nature by Design: A Teachers' Guide to Practical Nature Conservation* Birmingham: Urban Wildlife Group; Cantrell, R. (1989) 'Why not build your own nature reserve?' *Primary Science Review* No. 11, pp. 12–13; Mares, C. and Stephenson, R. (1988) *Inside: Outside* Brighton: Tidy Britain Group Schools Research Project/Brighton Polytechnic; and Young, K. (1990) *Using School Grounds as an Educational Resource* Winchester: Learning Through Landscapes Trust.
6   Greeves, J. (1990) 'Noise, dirt and litter – part of our everyday lives' *Questions* Vol. 2, Issue 4, pp. 6–8.
7   Mares and Stephenson, op. cit.

Chapter 5

Systems

Young children should be able to describe existing simple systems and to learn that a system is made of related parts with special jobs which are combined for a purpose. More able early-years and Key Stage 2 children should know that systems have inputs, processes and outputs and recognise these in simple systems. They can also be helped to appreciate that people or groups may have specialised roles in organisational systems. The children should have the opportunity to make simple systems, to evaluate and improve them and to use this knowledge to inform their design and making activities.

THE CHARACTERISTICS OF A SYSTEM

Many children are puzzled to find that some artifacts and environments can also be described as systems. It is important to realise that a system is only a generalisation intended to simplify and explain how things work. A map is used in a similar way to give a diagrammatic representation of a place to make it easier to understand. The concept of a system is not only used in technology, it is also used in many fields to attempt to explain very complex phenomena. The children may come across references to ecosystems which analyse the interrelationship of animals and plants with their environment; weather and river systems; social systems, interpreting how families, factories or political parties interact; and mechanical systems which explain how machines and their parts work together. Most of these are very complex, but more simple examples that children might investigate include bicycles, classrooms, schools, farms, supermarkets, libraries, orchestras and factories.

In order to help children to identify and evaluate a system the teacher needs to help the children to understand its basic characteristics:

- A system is a collection of components or elements which are related or linked for a purpose (that could not be achieved by the individual components by themselves).
- It has inputs, throughputs and outputs.
- Small systems (or sub-systems) may be linked to form more complex

systems.
- A system should have clearly defined boundaries.

1. A system is a collection of components or elements which are related or linked for a purpose

A heap of loose handlebars, brake levers, cables, brake blocks, etc. which could make the braking system of a bicycle is not a system until assembled because it lacks structure. Once made, these components are interrelated in special ways so they can do a specific task, i.e. transfer the pull on the brake lever so that the brake blocks squeeze together, thus stopping the bicycle. In the same way a sound-box, frets, strings, tuning keys and bridge can be assembled to make a guitar with the purpose of creating and amplifying sounds.

The components of a system do not have to be artifacts. They can be people, or even other systems (usually called sub-systems). The staffing system of a school may consist of headteacher, deputy head, teachers with special responsibilities, classroom assistants, nursery nurses, secretary, caretaker and so on, who all have specific tasks and interrelate to provide an effective education for the children in the school. A bicycle consists of several sub-systems including the braking and gearing systems. Other systems, such as an orchestra, may comprise artifacts and people.

When children make or examine products it is important to help them identify the components as this will stand them in good stead when the concept of a system is introduced. Use of manufactured constructional materials, such as Technical Lego, is also valuable as they help the children to appreciate that the different pieces or components need to be put together in a special way before the whole functions as one working model.

2. A system has inputs and outputs

There are three main parts to a system: input, throughput, and output or product. (See Figure 5.1.) The output is the way an identified need is satisfied. A guitar is intended to produce an aesthetically pleasing, controlled range of amplified sounds. (See Figure 5.2a.) The bicycle is one way of providing a

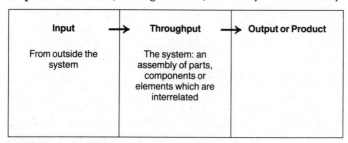

Figure 5.1 Main parts of a system

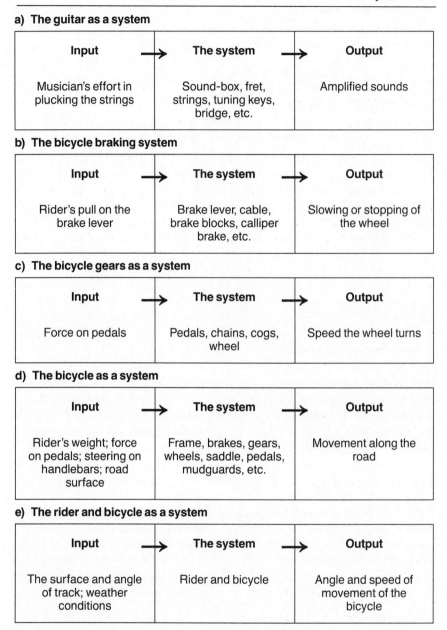

a) The guitar as a system

Input	The system	Output
Musician's effort in plucking the strings	Sound-box, fret, strings, tuning keys, bridge, etc.	Amplified sounds

b) The bicycle braking system

Input	The system	Output
Rider's pull on the brake lever	Brake lever, cable, brake blocks, calliper brake, etc.	Slowing or stopping of the wheel

c) The bicycle gears as a system

Input	The system	Output
Force on pedals	Pedals, chains, cogs, wheel	Speed the wheel turns

d) The bicycle as a system

Input	The system	Output
Rider's weight; force on pedals; steering on handlebars; road surface	Frame, brakes, gears, wheels, saddle, pedals, mudguards, etc.	Movement along the road

e) The rider and bicycle as a system

Input	The system	Output
The surface and angle of track; weather conditions	Rider and bicycle	Angle and speed of movement of the bicycle

Figure 5.2 Examples of common systems

faster and more comfortable means of travelling than walking (see Figure 5.2b). Whenever the children evaluate their own and manufactured products it is essential that the output is identified so they can assess how well the original need has been satisfied. Additionally this exercise prepares children

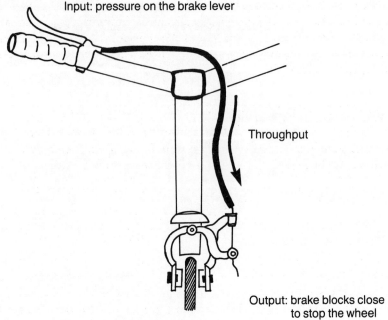

Figure 5.3 Braking system of a bicycle

to understand such products as systems.

Inputs are needed before the system can function. For example the guitar and braking system need human effort. (See Figures 5.2a, 5.2b and 5.3.) The inputs can also be other forms of energy, materials or even information. Machine systems need energy such as heat, electricity or human effort. A heating system or cooker can only work once the gas or electricity has been supplied. Factory systems need inputs of raw materials, and the staff in a school need instructions from the Department for Education, their local authority and governors as well as salaries! The system therefore may have several inputs. The whole bicycle can also be described as a system where the rider provides several inputs, including putting weight on the saddle; applying force to the pedals; and steering the handlebars. (See Figure 5.2d.) Again when evaluating products identifying the inputs will help children to understand systems.

3. A system has throughputs

Within the system there is not only an interrelationship between the components but also some transfer of 'material' between the units. The braking system transfers movement from the brake lever to the brake blocks. Other systems have other throughputs: food is conveyed through the digestive system; a central heating system moves heat through the pipes; parts are assembled to make a car as it passes along a production line; and instruction, ideas and knowledge are transmitted in the staffing organisation of a school. Identifying the sequence of this movement can assist children to see the interrelationships of a system. Indeed an ability to sequence is a prerequisite to describing and understanding systems.

4. Small systems may be linked to form more complex systems

In complex systems it is possible to find sub-systems, each of which has its own internal structure, inputs and outputs. It can already be seen that the bicycle could be regarded as having several sub-systems including:

The braking system	with brake lever, cable, calliper brake and brake blocks
The gearing system	with pedals, chain and gears
The lighting system	with bulbs, holders, wire and dynamo.

A system for making cakes might include the cooker as one sub-system with inputs of electricity and output of heat, and the cook and equipment as a second sub-system with inputs of ingredients and output of uncooked cakes. More complex systems may have more than one output as in the factory producing more than one product, or in a school which provides different facilities for children, parents and the community.

5. The boundaries of a system need to be defined

Each system should be distinguishable from the rest of the world by its appearance and behaviour so that it is clear where the inputs enter the system and the outputs leave. However, as the concept of a system is just a device for explaining how components in a whole interrelate, the boundaries of the system can be placed in the most logical position. In practice the decision where to place the boundary is often self-evident, as in the skin of the biosystem of the human body; walls around a shop; and the fence around a school or farm. Other boundaries are less obvious.

If the bicycle is considered as a machine the rider would not normally be thought of as a part of the system. However, if it was important to consider the dynamic balance of the bicycle when cornering at speed on a racetrack, it would be necessary to think of the person actually on it. In this case the input is the effect of the angle and surface of the track and the output whether the bicycle and rider stay upright.[1] On another occasion the main concern may be only with improving the gearing mechanism so the system chosen would be the interrelated cogs, wheels and chain. (see Figures 5.2c and 5.2e.) It is important therefore to be clear about the definition or boundary of the system being considered.

Merits of using the concept of a system

It is fairly easy to create a product that satisfies a need. However it is far more difficult to achieve this in an efficient and economical way. The idea of a system helps. By identifying the inputs and how the components relate it is possible to explore the effect of changing:

1. *The components:* For example once the basic bicycle has been designed it is possible to investigate what would happen if the size of the chain wheel were changed. Would it make the back wheel turn faster (output)? Would the rider need to provide more or less effort on the pedals (input)? Would the other components need to be changed and how? What would be the effect of providing a motor? Would the components need to be strengthened, redesigned or even totally altered? Would additional inputs such as petrol be required and what are the cost implications?

2. *The inputs:* The effect of changing the inputs can also be explored. Providing different types of batteries in a torch will affect the output in terms of the brightness of the bulb and time the light lasts. Such a change may also necessitate changes in the components such as the size of housing and type of bulb.

This process enables the technologist to improve the design of the parts and the whole product.

SYSTEMS VARY IN SEVERAL MAIN WAYS

When introducing the idea of systems to children it is important to start with simple, small-scale, closed systems where the children can reasonably predict the outcome and can change the inputs and components so that they can explore and understand their effects.

1. *Complexity:* Some systems have many elements or component parts even if the output is similar. A motor car and bicycle have more or less the same functions but the car has more components and is more complicated.

2. *Scale:* The gearing mechanism, bicycle, bicycle shop, and bicycle manufacturing industry are all systems at a different scale and can be seen as sub-systems within bigger systems. Systems can be defined at all extremes of scale, from a hand drill to the solar system which explains the interrelationships of planets and the sun. In turn the solar system is also only a minute sub-system in a galactic system.

3. *Certainty of output:* In some machines such as the bicycle the interrelationships can be described relatively easily and the output can be reasonably predicted. In others which are more complicated or as soon as people are involved, the outcome is less certain. It is doubtful that all the nursery children will perform as planned in their assembly, or that the profits of a complex national industry can be accurately anticipated.

4. *Closed or open:* Some systems like the bicycle are relatively 'closed': it is

clear which components belong to the system and there is little dependency on outside factors. Other systems are more 'open'.

DEVELOPING CHILDREN'S ABILITIES TO DESCRIBE, EVALUATE AND DESIGN SYSTEMS

Children can study systems in many different contexts, such as the home, school, recreation, community, and business and industry. Although children of all ages can investigate systems in all these areas, the early-years child will need to concentrate on very familiar situations and simple systems. As the children mature they can study systems that are unfamiliar and more complex. A possible progression for developing children's ability to describe, evaluate and design a wide range of systems might be:

1 Sequencing activities to recognise the succession of processes from input to output.
2 Investigating simple machines which have easily identified components.
3 Examining and devising simple organisations and systems in the school and locality.
4 Investigating and evaluating systems of recreational activities.
5 Analysing and designing business systems.

SEQUENCING

In order to describe a system children have to be able to recognise the sequence of a series of processes from input to output. Later, as they design their own system, they will need to be able to draw up a succession of activities before starting a project. Firstly, therefore, children require activities to sort out the correct sequences of activities they have already performed, which they can record in writing or drawings. Initially these tasks should be very simple with few steps. Later the children can record more complex undertakings. These experiences should subsequently help them to plan a series of activities before starting a project.

Many teachers already include sequencing activities in language and mathematical work, and these might include ordering number, letters, shapes and sizes of different items. For technology sequencing a series of actions is particularly important as the children need to see the importance of planning a process. The children could be given pictures and asked to sequence them into a logical order or be asked to retell a familiar story either orally, in pictures or in written form. They might also order their activities during the day which can be recorded in the form of drawings, comic strips, mimes, photographs or typical sounds of each activity.

Once the children have a good grasp of sequencing they can start to relate these to systems. They might look at particular class activities in more detail, such as the stages in painting a group picture, getting ready for an apparatus PE session, growing seeds, making a book, and cooking biscuits. After a cooking activity the children can recap the stages and record them in some way. The less able or very young children may need to be given pictures showing the stages in order to sequence them in the correct order. The children should be able to see that the steps need to be repeated if the biscuits are cooked again and that the order is important. There would be little point in washing hands after handling the food, and impossible to put the mixture onto the baking trays before mixing the ingredients. Once the stages have been identified the teacher can point out that the children have produced a simple system with inputs (ingredients) and outputs (cake).

As the children become more competent they could produce a simple flow chart for familiar activities, such as making a drink of orange or a cup of tea and wrapping a parcel, which they can try out by asking other children to follow their instructions. The tasks can become increasingly complicated such as drawing up instructions for making games, constructing models and controlling traffic lights.

SIMPLE MECHANICAL SYSTEMS

The idea of systems can be introduced to young children during the scientific investigations of toys and household machines and when making moving artifacts. The children can handle, or ideally build, the parts, in order to see that the machine is made of components which fit together. As energy is the main input for machines the children might also be presented with the idea of inputs and outputs.

The following steps can help children to describe each system:

1 Clarify who it is intended for.
2 Identify the end product. What is the purpose of the machine or system?
3 Identify the elements, component parts or sub-systems.
4 Identify the inputs. What is needed to make the machine or system work?
6 Follow the sequence of action from the input to the output.
7 Try to work out the relationship between the parts. An annotated picture or diagram may help.

For example a school hand drill might be investigated. Its main purpose is to make drilling holes easier for children. It has several components each with a specific role: the hand grip made of wood for holding the machine in place and to press downwards on; the handle to enable effort to be applied in a comfortable position; a series of gears to increase the speed and change the direction of the force; the device for gripping and changing the bit; and the bit itself with a spiral cutting edge. The main input is force and energy provided by a child's effort and the output is the turning motion of the bit. A simple drawing with the direction of movement of each gear marked will help to show the transfer of energy through the machine.

Once the children have described the system the following steps can help to guide children to evaluate it and suggest improvements:

1 Speculate what happens to the other components when one is changed.
2 Consider what might happen if the inputs change.
3 Consider what works well and what parts do not.
4 Suggest some improvements. Could it be simpler or more efficient in some way? Could it be made easier to use, cost less or look more attractive?
5 Identify what components or inputs need to be changed to achieve the suggested improvements.
6 If possible try out the improvements and evaluate them.

Once the children have described the drill, they might consider the effect of changing the shape or size of different parts, such as the gears, bit or handle and what happens to the machine and output if the speed the handle is turned varies. They may be able to make suggestions for improvements. They might suggest giving the handle a more comfortable grip, adding an electric motor or making the grip for the bit easier to open. Some suggestions, as in the design of the handle, could be done with minimal effect on the whole system, others would have more far reaching effects. It is often difficult for children to construct their ideas so it may help to look at other machines of the same type in order to compare them with the original. (Simple machines are examined in detail in Chapter 6.)

EXAMINING AND DEVISING SIMPLE ORGANISATIONAL SYSTEMS IN THE CLASSROOM

Once older early-years or young junior children have a good grasp of simple mechanical systems they can investigate organisational systems. These are more difficult for children to describe as the components and their interrelationships are not always clear. However they should be able to identify the inputs, outputs and sequence between the two. Children need to be able to describe and evaluate a system before they can effectively design one doing a similar task. The children might first carry out an activity which the teacher knows can be described as a system. This could be giving a small group of children the task of arranging the items in a cupboard or producing a puppet show which the children are allowed to complete by trial and error. After finishing, the task can be sequenced and the components of the system identified so that the children can be shown that their project could be described as a system and improvements suggested. A similar activity could then be repeated but this time planned as a system before starting.

Organisational systems can involve arranging artifacts in a logical way so that there is a special place for each item as in a filing system or book in a library. The organisation on the other hand might be a way of grouping people to perform a task efficiently. Other organisations as in the orchestra involve both artifacts and people.

Organising artifacts

The organisation of equipment can be analysed as a system. In this case the input is new materials or equipment coming into the school and the output is tidy, accessible items that can be easily cared for, used and monitored. The system itself is the storage compartments, locations and the ways the materials are sorted and controlled. A filing system might consist of a cabinet with labelled dividers and spaces for files in alphabetical order, and within each file the papers may be ordered according to date. There is then a clear boundary with relationships between all the components. Depending on the organis-ation in the school there may be opportunities for children to look at the way PE equipment, woodwork tools, paper, pencils, paints and other consumable materials are stored in their classroom or school.

One class of year 3–4 children was given the task of designing the set-up of a walk-in cupboard which was to house all the materials and tools they might need for art, craft and technology. Initially small groups worked on the problem and drew out their design. These were shared and a plan agreed. The children first needed to be clear in their own minds who was expected to use it and what would be stored there. They tried to think of a logical sorting system and appropriate containers and ways of enabling users to find what they wanted quickly. The children quickly realised, as they were the main users and they could not reach or even easily see the top shelves, they would have to put the frequently used equipment on the lower shelves. Heavy, very bulky equipment was kept on the floor and labels put on the shelves in appropriate places. Materials of the same type were stored together, i.e. a special location was given for painting materials, drawing and colouring, collage, woodwork and so on. They decided each item should be kept in separate transparent or open boxes clearly marked with pictures and with English and Punjabi labels (most children spoke both languages) so it was easy to find what was wanted and no one had an excuse for just throwing items into the cupboard. The plan was evaluated as they went along and adjustments were made to take into account availability of containers. The children also realised eventually that some type of on-going monitoring system was needed to restock and keep the area tidy.

Organising people effectively

Persuading children to keep the classroom area tidy is often a problem. By encouraging the children to evaluate the system of tidying up, everyone might be helped. At the end of tidying up a particularly messy session the children could be asked to think about what had occurred and what tasks they did, and produce a diagram of what happened. Of course if the teacher has a video, and the children are used to having it used in the classroom, their actions can be recorded. Their discussions should lead to suggestions for improvements. Did everyone do something? Did everyone do a fair share? Did particular people do certain jobs? Did some children take the role of the leader? Was this necessary? Were all the tasks done satisfactorily? Was anything left undone? How long did it take? The children could then be asked to consider the task as a system with the input being the mess and the output a tidy classroom, and consider how they could organise themselves into groups with particular responsibilities and in what order the tasks need to be done. The class, groups or individuals could draw up different system designs, given a similar mess, that would be more efficient. The systems could be tried out after subsequent messy sessions and evaluated. This system is principally the organisation of people to do specific tasks in a particular order.

Organising people and artifacts

The preparation for a class assembly, or indeed the school play, can be described as a system involving both people and artifacts. The inputs are the materials, equipment and story to be acted, and the output is the performance. The actual system is rather complicated as it usually involves several sub-systems, such as making the props, preparing accompanying music and rehearsing the play. However the children can identify the inputs, outputs and the sequence of activities to achieve the final product. The experience can lead to them seeing the importance of planning to time their project carefully.

Giving small groups of children responsibility to produce and analyse some simple production for themselves should enable them to apply the experience to a more complicated class assembly with more people involved. One class of year 4 children were put into small groups with the task of producing puppet plays based on familiar traditional stories. The children worked mainly by trial and error but at the end were able to identify the sequence of activities that had made up their task. One group of five children produced the 'Three Billy Goats Gruff'. Their main steps were: constructing four puppets; making scenery; deciding what the narrator and characters would say; trying out the play; adding sound effects and music; inviting the rest of the class; and finally the performance. Subsequent groups found they used a very similar sequence of activities which could be used to plan a performance before starting. The class, with support, were then able to plan a system for a whole class assembly giving different people specific roles which included acting, making scenery and costumes, and producing accompanying music; invitations to parents and other classes; and providing biscuits and drinks for parents. All these could be planned beforehand with a simple sequence or timetable to fit in with times the hall can be used for practices. In the past the class teacher has made all these decisions, but by involving the children they are often more tolerant of the need to wait while others practise their parts, and it maximises the learning opportunity provided by the assembly.

Many technological activities can be planned by using the concept of a system. For example a junior class might be divided into groups and given the task of making their own musical instruments in order to produce a piece of music to accompany an advertisement for a fizzy drink, to accompany a short film about a haunted house, or for background music in a clothes shop. By giving a fairly short time for the whole task the children need to share the tasks. If they have had experience of evaluating and designing systems they may be able to apply the idea to carry out this task more effectively. The children could also be involved in drawing up the organisation of sports days, fund-raising fairs or mini-marathon runs and book weeks.

SYSTEMS IN THE SCHOOL

Children need to describe fairly simple and familiar systems that are commonly found in their homes and schools before trying to suggest improvements. The subsequent experience in improving existing systems should help them to plan or set up similar ones. The systems produced might be in the form of diagrams, models or actual examples.

There are several systems both mechanical and organisational that can be studied in the whole school, including heating and security systems; organisation of lunch-time supervision, cleaning the school and preparation of dinners; and administrative systems such as collecting and recording dinner money.

School heating system

The purpose or output of a central-heating system is to provide equable heating throughout the school, and the components include the boiler, pipes, water and radiators. When the children investigate the effect of heating and cooling substances as part of science they could also examine this system at the same time. In many schools the heating system and the security system are also examples of automatic control.

Security systems

The children may not be so aware of the security system, so they can be asked how they think the school is kept safe when no one is in the building. They could also compare this system with that in homes and collect information about how to make homes safe. There will be very clear boundaries in such a system which may have several features to protect it such as outside lights that automatically come on in the late evening, different types of locks, and toughened glass or wire meshed windows. The inside of the building may also be protected by several sensitive beams carefully located throughout the area that start an alarm when broken, and there may also be specially protected areas such as a safe and store for computer equipment.

After having studied existing security systems the children might make and design some of their own for a specific area. The working devices might include some sort of pressure pad which can be made by using pieces of aluminium foil on a folded piece of card. When someone enters the area and steps on the pad, the card is pressed together and a circuit completed enabling an alarm to sound. Another method is by using a light beam which is broken as someone enters and sounds an alarm.[2]

Provision of school dinners

If the staff are amenable, the school kitchen can be studied to see how the food is ordered, menus decided upon, food prepared by different staff, organisation timed to have all the different dishes ready at the correct time, tables prepared, children served, and dishes cleared and cleaned. Some of the children may be part of this system if they help with the clearing up and may be able to suggest ways of improving it. The school kitchen is a complex system which depends on careful timing and control of stock, and children will appreciate how serious it would be if their meals were not ready in time or there was too little food.

The school library

Libraries have great potential to be studied as collections of artifacts including books, seating arrangements and display facilities; as environments to promote interest in and facilitate the use of books; and as systems. The school library is an accessible system which acts as an introduction to more complex libraries that the children may be able to visit in their community. As a first step the children could describe their school library. As with all systems they need to be clear about what is the aim of the system and who is it for. The teacher may need to ask questions to prompt the children's thinking. Who uses the library? What are their ages and reading abilities? Is the library only for choosing books or is it used for other purposes, such as reading, story-telling or a classroom? Most libraries are intended to enable the maximum number of people possible to have fair and effective use of the available books.

The children should also try to identify all the component parts of the system and work out what their functions are. As it is a complex situation it is helpful to:

1 Draw a map of the library with the constituent parts marked and annotated with the purpose of each feature. (Such a map could also be used

to discuss the environment where the emphasis would be more on the significance of the relative positions of the facilities.)

2 Write down the sequence of choosing, and then later returning a book.

Once the children have looked carefully at the library they can be helped to describe it as a system. Groups of children could investigate two overlapping sub-systems:

  1. *Organising and storing books:* This includes the shelves and book boxes; arrangement of books into alphabetical order, subject, reading level, book type, fiction and non-fiction; as well as labels and card indexes.

  2. *Helping people to use the library fairly and effectively:* This may involve the use of library tickets for the user and for each book; comfortable seating so time is spent choosing; information in the form of posters and booklets about using the library and new books; indexes for finding books; and people available to help.

After the children have described and analysed their library they may be able to suggest ways of improving it as well as being more aware of how to use it competently themselves.

Organisation of staff

The study of the responsibilities of the adults in the school can be a useful way of introducing the concept of a hierarchy, which is very common in many organisations. The children could interview and draw pictures of all the staff and then try to arrange the pictures to explain how the school works. Often the children will suggest the headteacher's picture goes at the top but they may equally suggest the secretary or caretaker.

SYSTEMS IN RECREATIONAL ACTIVITIES

The orchestra

Investigating how individual musical instruments produce variations in pitch, loudness and timbre are important elements in the science curriculum.[3] Children also learn to make music for themselves and write simple diagrammatic scores. Additionally as part of music activities children are often introduced to different instruments of the orchestra and some fortunate schools are able to have visits from orchestral players who talk about their instruments and music. These experiences can be related to the study of the orchestra as a system that incorporates both artifacts and people to produce one end product, i.e. a piece of music. The children should discover that each person has a specific and specialised role which can only function effectively in combination with the other elements of the system.

The modern symphony orchestra usually has about eighty players with instruments of all sizes which are divided into four groups or sub-systems:

string, woodwind, brass and percussion. In the string section four instruments are used, usually about twenty violins, eight violas, six cellos and four double basses. The main instruments in the woodwind section are the flute, clarinet, oboe and bassoon, and there are commonly only two of each. The brass family often includes three trumpets, three trombones, four horns and one tuba, and finally the percussion family consists of instruments that are struck, such as the kettle drums, tambourine, triangle and cymbals. The whole system therefore is made up of four different-sized groups of people who play in unison to produce a sound. The range and type of sound produced by the group are controlled by the design of their instruments. The action of the system is planned by the composer and directed during the performance by the score and conductor. The orchestra is an excellent example showing the interrelationship of the parts which is effected by the quality of the performance of each part, both personnel and equipment.

The swimming pool

The organisation of many sporting venues and activities can also be described as systems. If the children visit the local swimming pool they might also consider this system which comprises two main sub-systems:

1. *Personnel organisation:* This probably includes a manager, instructors, attendants, engineers and reception cashier.

2. *Pool and heating mechanisms:* These incorporate the boilers, filtering and cleaning plant.

The children can also investigate other sporting and leisure enterprises such as football clubs, sports centres and ice rinks.

EXAMINING AND DESIGNING BUSINESS SYSTEMS

As Key Stage 2 children become more confident with the idea of a system, unfamiliar and more complex ones can be investigated. Their diagrammatic representation of these should also become more sophisticated and might be set out in the form of annotated diagrams, flow charts and plans.

The shop

As the shop can be seen as an environment, a system and a collection of artifacts, the teacher may well wish to explore all these facets in one project, particularly as shops are very familiar to children and there are opportunities to set up role-play shops in the classroom or to actually run their own shop in the school. When discussing the purpose of the shop children tend to assume it is a service for them and their families, but as they discuss it further they should be able to appreciate that its prime purpose is to provide the maximum possible profit for the shopkeeper or company.

Once the output of the system is clear it is easier to understand why the different components are necessary. The 'boundaries' of the system can be fairly easily seen as the actual shop building. The inputs are the goods and output the profit. The system will aim to use the work force economically; have a well-organised method of delivery and display of needed goods; use ways of encouraging the customers to buy as many items as possible; have an efficient way of collecting payments with minimum errors and loss of money; and have an effective security system so that goods or money are not stolen.

Initially the children need to collect information about the components which are composed of both physical elements and personnel. If the children were studying a supermarket they might start with identifying the main physical components on an overall plan of the shop, such as the shelves, cash desks, trolley parks, freezer cabinets and delivery bay, and try to explain what their functions are. Drawing up sequences of different actions can help the children to appreciate how the components interrelate. The sequence of what happens from delivery of a batch of perishable items, such as yoghurt, could be outlined until the item leaves the shop, either with a customer or to be disposed of because it is out of date. They might also follow the actions of customers from the moment they arrive in the environs of the shop to buy various items to the point of departure. The children could also list the personnel and their roles and even draw up a simple chain of command.

As the whole system is so complicated the teacher needs to help the children identify the different sub-systems, some of which could then be investigated in more detail. These might include:

1. *Organisation of goods:* This might include examining the different shelves and display cabinets; how different types of goods are grouped and set out; what notices are provided to guide the customers; and how the goods are date labelled and priced. This sub-system ensures that there is a special location for every item so that they can be found and monitored.

2. *Ordering and delivery:* The manager must know which goods are being sold and need to be restocked. This may involve looking to see if the shelves are becoming empty or recording each item as it is sold. Once it is known an item is needed the children could investigate how the manager tells the supplier, and how it is delivered, moved to the shelves and priced.

3. *Checkout:* In order to collect the money for each item the goods have to be handled individually in some way by the cashier so that the price can be identified and recorded. This might be done by mental arithmetic, recording on a till or using bar codes. The goods are packed either by shop staff or by the customer, and payment is made in various ways and the cash put in a secure place. In order to move people through the system quickly and keep customers happy with the shop there may be ways of sorting customers by making special arrangements for those with cash or very few items, or for disabled people.

4. *Security:* There are systems for security when the shop is open and when

it is closed, which may entail closed-circuit televisions, security staff, alarm systems and methods of collecting and storing money which can be examined by the children.

5. Organisation of staff: In any large organisation the different tasks are identified so that people can be trained and are clear about their responsibilities. There is usually a clear chain of command so that each person is not given conflicting instructions. The children can find out how many people are employed and their specific jobs, and who is in charge of the different areas.

By investigating one shop in detail the children should be able to apply the experience to different situations. They might try to design a system for different types of shops, such as sweet shops, newsagents, clothes shops and/or shoe shops, which could then be used to set up a role-play situation, perhaps for younger children. They might also set up their own shop to sell snacks or they could grow and sell plants to raise school funds. Based on their experience of studying existing shop sub-systems the children can ask themselves questions in order to guide them in designing their own: What should they sell? How do they buy in what they need and store it? Who should run it and who is responsible for decisions and care of goods and cash? How will they protect their goods and money received? How will they inform, and cater for the needs of, their customers?

There are many other business and industrial systems the children might investigate in the community, some of which demonstrate the importance of sub-dividing tasks as in a factory production line while others, like a café or fast-food outlet, show the significance of timing and control of stock. Enterprises and organisations, including farms, sewage disposal, building and concrete industries, that will be studied in other subjects are especially worth investigating.

A catering enterprise

The children could set up a café for parents or visitors at a school event or celebration, or they might make the arrangements for the end-of-term school party. Different groups can carry out the detailed planning for the various tasks or sub-systems: obtaining and preparing food and drink; advertising the occasion; arranging the decorations, furniture and equipment and clearing the site; providing entertainment; and organising the distribution of food and drink at the event. Their plans can be evaluated and improved by the whole class, a timetable drawn up, and tasks shared out. This type of practical experience will enable the children to become more aware of the significant features if they visit a café or fast-food store. Alternatively a visit by younger children to such an enterprise could lead them into making a role-play café which includes the main components of the system they have visited.

The farm

The boundaries around the farm give a fairly clear delineation of a system and the inputs and outputs are fairly obvious. The inputs for a dairy farm, for example, include the pasture seeds, fertilisers, weed-killers and livestock and the output is the milk. However, as with all complex systems, it is sometimes difficult for the children to identify all the parts and then to express their

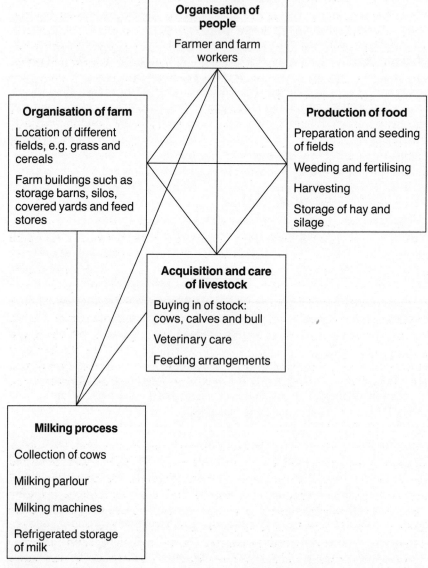

Figure 5.4 Simplified diagram showing some of the interrelationships between sub-systems on a farm

findings in some way. After visiting a dairy farm the children could try to list all the components of the system and then group them in some way related to their main purpose. Where the interrelationships are not obviously progressive a diagram with lines drawn in to show relationships might be more appropriate than a sequential list. (See Figure 5.4.)

The sub-systems can be studied in greater depth. The milking parlour in particular has a lot of potential for technological projects. If the children are able to watch one working they can sequence the actions taken by the dairy worker and identify the devices that enable the whole system to run efficiently. Later they might try to make models of some of these, such as a gate which might be a simple lifting barrier activated by a pressure pad in front of it that only gives access to a limited number of cows, and/or a machine that automatically detects when the cow arrives at the station and delivers a measured amount of food into a trough.

It must be remembered that the dairy farm is only part of a bigger system designed to provide milk on the doorstep, which involves timetabled tanker collections, testing laboratories, the creamery, milk depot and delivery system by the milkman.

Water and sewage treatment

As part of science and work on environmental issues children may study how water is treated. This can be extended into examining how water treatment can be described as a system. The actual system is fairly simple and has a clear sequence of processes. There are a limited number of inputs, i.e. dirty water and chemicals to clean and improve the quality of the water. The output is clean water that is both safe and desirable to drink. The project can also enable the children to discuss the effect of having incorrect inputs and the consequence this has on the whole system. Sewage treatment has a similar process of filtering and settling and can be described as a system in a similar way.[4] The children may also be able to appreciate that treatment works are only part of a larger system in each Area Water Board for caring for water supplies and delivery to homes.

The concrete industry

The concrete industry is an ideal study for primary classes, both for science and as a system, because the basic raw materials and end product are within the common experience of most primary children; the actual industrial process is relatively simple to understand; and it can be brought to life by many related practical science activities. Ready-mixed concrete plants are likely to be fairly local because the mixing process must be carried out near to potential customers to avoid the concrete hardening and becoming unusable during transportation, and most schools are also close to building sites where the end product can be seen to arrive and be used.[5]

Input: raw materials – sand, gravel, clay, limestone

Sub-system:

Cement production
- Extraction of clay and limestone
- Transport of clay and limestone to cement plant
- Production of clinker
- Clinker ground into cement

Sub-system:

Extraction and sorting of aggregates
- Geological survey
- Planning permission for extraction
- Extraction of sand and gravel

Sub-system:

Plant for sorting and grading
- Removal of undesirable materials, e.g. lignite
- Gravel and sand washed
- Gravel sieved and graded
- Silt removed
- Sand sorted and graded

Sub-system:

Concrete plant
- Storage of aggregates and cement in hoppers
- Orders received from building contractors
- Materials weighed and mixed in truckmixer
- Transport to building site

Sub-system:

Testing and quality control

Figure 5.5 Simplified system of the ready-mixed concrete industry

There are four main parts or sub-systems of the concrete industry: cement manufacture, extraction and sorting of the aggregates, mixture of these two elements to make concrete, and a testing programme throughout to maintain the quality of the product. Within these systems are other sub-systems as can be seen in Figure 5.5.

Once the teacher and the children are familiar with the concept of a system they should find many other examples around them. Others that can be easily investigated include travel agencies, railway stations and airports, multistorey car parks, and the organisation at a hospital or health centre. Such experiences will provide a valuable foundation for describing and understanding many other complex organisations such as national and international industrial and economic phenomena that will be covered in later geographical studies and in biological environmental interrelationships. It also provides an approach that should help people to organise their own technological and other activities in a more planned and logical way, particularly when working with a large number of people.

NOTES AND REFERENCES

1 Romiszowski, A. (ed.) (1970) *The Systems Approach to Education and Training* London: Kogan Page.
2 This can be set up with a few small items from kits fitted together with batteries, e.g. ALPHA electronics kit from Unilab light sensor (unit 11), transducer driver (unit 40) and buzzer (unit 31).
3 Programmes of Study for Key Stages 1 and 2. DES (1991) *Science in the National Curriculum* London: DES and Welsh Office/HMSO.
4 Mares, C., Blanchard, H., Stephenson, R. and Redhead, M. (1988) *Our Environment: Teacher Guide* Walton-on-Thames: Thomas Nelson has details on water and sewage treatments and related experiments.
5 Jarvis, T. (1992) 'Making science relevant: Investigating the concrete industry' *Primary Science Review* No. 24, pp 16–19 gives further classroom activities based on the concrete industry.

FURTHER READING

Bale, J. (1976) *The Location of Manufacturing Industry* Edinburgh: Oliver & Boyd.
Dury, G. (1981) *An Introduction to Environmental Systems* London: Heinemann.
Ross, A. (ed.) (1990) *Economic and Industrial Awareness in the Primary School* London: PNL Press.
Smith, D (ed.) (1988) *Industry in the Primary School Curriculum: Principles and Practice* Lewes: Falmer Press.
Technology for Teachers Course Team (1975) *Fundamental Concepts in Technology II: Units 5 and 6 Control and Systems and Design* Milton Keynes: Open University Press.
Tidswell, V. (1976) *Pattern and Process in Human Geography* Slough: University Tutorial Press.
White, I., Mottershead, D. and Harrison, S. (1984) *Environmental Systems* London: George Allen & Unwin.

Chapter 6

Energy and machines

Young children should use a variety of energy devices to make things move, use simple mechanisms to transfer motion and recognise that a source of energy is required to make things work. They might do this through the study of toys which can be classified in terms of the energy needed to make them move: battery, muscle power, solar cells, elastic, spring, wind and so on.[1] Later in the primary school the children should have the opportunity to apply their knowledge of different energy sources in their designs to make devices such as simple model windmills, boats and land yachts.

In both technology and science, during Key Stage 2, children are also expected to learn about some of the basic types of movement and mechanisms to be found in machines, the most common being wheels and axles, pulleys, gears, cams, levers, wedges and screws; their function; and how they can be controlled. In addition, in technology, by the top of the primary school children should have had the opportunity to apply some of these ideas so that they can select and use simple mechanisms, including linkages and gearing, in making prototypes. They should be able to use mechanisms to change one type of motion into another and identify the basic principles of how mechanisms change speed or motion, from one form to another.

Once the children have an understanding of these mechanisms their studies can be extended by examining simple machines as systems. Many toys and basic household devices are very suitable for introducing the concept as they are simple, closed systems with fairly obvious inputs and outputs and a limited number of mechanisms or components. (See pages 64–5.)

HISTORICAL DEVELOPMENT OF MECHANISMS AND MACHINES

Children need to appreciate that the process of technological development has always been prompted by people's need or desire for better conditions and that it is accelerating because more devices and ways of using energy have already been invented and are available for further innovations. A project on machines not only enables the children to investigate the basic mechanics necessary in science and technology, it also allows them to review the development of those machines historically.

In retrospect early technological development seems to have been painfully slow. For centuries the energy for work was provided by humans and other animals, with a few simple machines to help them. The pyramids were built using levers and the power supplied by men and oxen. If the children consider the lifestyle of early people they will appreciate why the first machines made pots (3500 BC), wove cloth (3000 BC), pressed olives and grapes (1600 BC), measured time (1500 BC) and lifted water (800 BC), as these were identified as the most important needs of the time. By AD 285 Pappus of Alexandria described five useful machines: a cog, lever, pulley, screw and wedge.[2] Once these fundamental devices were invented they formed the basis for more complex machines. Indeed most modern mechanical devices consist of one or more of these basic simple machines.

The relationship between perceived need and technological development can also be seen when considering where different innovations occurred. The supply of water for irrigation, drinking, and domestic and industrial use was significantly more difficult in most Muslim lands which did not have the easy access to water compared to Northern Europe's heavy rainfall and abundant rivers. Consequently many inventions to raise water came from the Muslim area. The shaduf is rather like a seesaw with the pivot about two-thirds along a beam. A bucket is fixed at one end and a counterbalance made of stone or clay is attached to the shorter end. Another machine invented in the area and called the saqiya uses two large wooden gear wheels so that they mesh together at right angles. As an animal turns one wheel, the other turns and raises the water which is collected in pots attached to the wheel.[3] By looking at inventions in different cultures the link between the need of the community and imaginative responses can be explored and the children are helped to appreciate that the western world does not have a monopoly of creative technological development.

Ways of making the use of energy more efficient led to other significant technological changes which had inevitable effects on society which then created different problems and needs. Sail was the first mechanism to harness natural power and windmills originated in Persia about AD 500. Water power was developed in the Near East a few years before Christ. These early water-wheels had a vertical axis with a horizontal wheel in the water which turned as the paddles caught the current. Early Vitruvian water mills (100 BC) had a different design and worked by the lower part of a vertical wheel being submerged and turned by the force of the water's flow. The Domesday Book (1086) reports 5,624 water-mills in use in Britain south of the River Trent. These mills had a great impact on industry and society as they were used for grinding flour, crushing ore, operating bellows and in saw mills.

In the seventeenth century machines were invented to use steam. The first were pumps to remove water from mines. Later pistons were used to turn wheels, enabling the development of steam trains and engines to drive other machinery. A study of the industrial revolution gives an example of the social upheaval that new technology can create and demonstrates the great misery that can be created by lack of concern for individual human needs in such development. Petrol-gas engines caused another significant change as did the discovery of electricity after Volta's battery cell (1800) and Faraday's electromagnetic motor (1821). Relatively recent inventions such as atomic energy, the jet engine and the development of electronics have again had far-reaching effects on society and its organisation. One of the important lessons from reviewing any sequence of technological development, but particularly that of developing machines and energy use, is that it is both dynamic and irreversible. Children need to appreciate the effects of technological development in the past so that they can anticipate and hopefully avoid excessively damaging consequences by considering the possible, far-reaching effects of contemporary inventions.

INVESTIGATING SIMPLE MACHINES

A machine can be considered to be a device that uses one form of energy, modifies it and delivers it in a more suitable form where it is needed. It makes work easier by either reducing the effort needed or making it easier to use the effort. Children usually assume that a machine is a complex device requiring electrical energy, whereas it can be extremely simple such as a tin opener, pair of scissors and hand whisk. The hand whisk, for example, moves energy supplied by the hand to the beaters; and by using a series of cogs, one turn of the handle on the whisk turns the beaters several times, making the task of whisking an egg far easier than using hands alone.

Initially children could be given simple household machines to discuss and try to work out how they work. Items such as a bottle opener, ice cream scoop, hand mixer, hand drill, old clocks, scissors, nutcracker, spade, claw hammer, wheel barrow and tweezers give a variety of types of machine. Once the children have had an opportunity to study the machines in detail they could sort them according to the energy source used, try to classify the movements and identify different types of mechanism.

There are four basic types of movement:

Linear: movement in a straight line between two points
Reciprocal: movement backwards and forwards in a straight line,
              as in a piston
Rotary: circular movement, such as a wheel or cog
Oscillating: backwards and forwards movement in an arc,
              e.g. a pendulum.

There are several basic mechanisms which form the basis of both simple and complex machines. These include wheels, axles, cams, pulleys, gears, wedges, screws and levers. Children can also design and make working models which will help them to identify and explore all these fundamental ideas. For example, following visits to a building site, the children could make model working cranes with winches and pulleys or they might construct fairground models, including a Ferris wheel and roundabout, with wheels, cams and gears. Construction kits are particularly useful for helping children experiment with different mechanisms and for working out their ideas.[4]

THE BICYCLE

After study of the different basic mechanisms in fairly simple devices, one machine and its development could be studied in detail. The modern bicycle, for example, is made of many different mechanisms. Bringing a bicycle in to the classroom not only excites the children's interest but also allows them to look at it closely. One way of drawing their attention to the various parts is to either give them a diagram of a bicycle or to ask them to draw one annotating it so that all the materials used are identified. It is important to go beyond the descriptive and ask why a particular part is designed in that way. The children will usually find that the saddle, pedals, handles and tyres are covered with plastic or rubber to provide comfort and increase friction to improve grip. The frame is usually made of hollow tubes of metal in basic triangular shapes to be both light and strong. Other components include air-filled tyres fixed to light metal wheels kept in shape by a triangular pattern of spokes; handlebars linked to the front wheel so direction can be controlled; pedals, chain and gears to efficiently use the rider's efforts; and some sort of lighting device. By looking at pictures of bicycles, perhaps from a catalogue, of children's tricycles, small bicycles with stabilisers, tandems and racing bicycles, this basic system can be identified in most modern bicycles with slight variations such as size, number of wheels and gears in order to cater for the needs of different users. The children can compare and evaluate the designs with a view to suggesting improvements so that they can realise that altering a basic existing system is the usual approach for making new technological improvements. There are very few totally original inventions or products.

By studying the historical development of bicycles this process of adapting an initial model either to cater for a new set of users or to improve the comfort and/or efficiency can also be seen. It may be possible to take the children to a museum which displays old bicycles or use slides so that the children can be prompted by open questions to try to find the stages of development for themselves and suggest why they occurred.

The hobbyhorse was first used in France over 100 years ago and must have been inspired by horses as the first examples had a replica of a horse's head. It was made of wood with two wheels, had no pedals and was walked along by the rider pushing his feet along the ground. Walking through the mud was obviously unsatisfactory and Kirkpatrick Macmillan in 1839 invented the first self-propelled bicycle that used two cranks which were pushed backwards and forwards by the feet to turn the wheels. The popular 'boneshaker' (1861) was made of wood and iron, improving the durability of the machine, and had pedals on the front wheel. However the machine was heavy, a lot of effort was needed keep the wheel turning, and the spoon brake device on the wheels was not very effective. These, then, were things that subsequent developers could see needed improving.

The penny-farthing (1870) was much lighter, as the wheels were made of iron with thin wire spokes, and required less effort to ride because of the very large front wheel. The pedal was fixed to the centre of the wheel so for every turn of the pedal the bicycle went forward the same distance as one circumference of the large wheel thus maximising the distance travelled for each rotation of the pedal. The penny-farthing is a good example of how the requirements of the user affects the proportion and size of a product. The leg measurement to reach the pedals was crucial as riders wanted to travel as far as possible when pedalling, but if the distance from seat to pedal was too great they would not be able to reach the pedals. Consequently each bicycle had to be made to fit. This design is also an example of how by solving one technological problem another can be created, as getting on and off the machine was rather hazardous. Consequently the chain-driven tricycle was designed to cater for older men and women, which was much safer and more stable but had no brakes. The rider had to attempt to stop by back-pedalling!

By 1885 the Rover Safety bicycle had been invented, which had many improvements and perhaps demonstrates how improvements accelerate once the basic system has been designed. These improvements included improved steering; a chain-driven rear wheel; pneumatic tyres which were lighter and cushioned the rider against bumps; seating placed between the wheels where there would be less vibration; and a spring seat which made the seat more comfortable. There is probably no ideal invention, and although this was a great improvement the lower positioning of the saddle resulted in the rider getting muddy. As different groups of people wanted to use bicycles, adaptations needed to be made to cater for their particular needs. For example a step-through frame was designed for ladies with long skirts and mudguards provided. Racers, however, wanted a more streamlined shape so the design of the handlebars was altered and lighter metal used, and mountain bicycles needed to be compact, strong and have many gears.

By looking at the sequence of developments the children can see how the process of technological change builds on past ideas in response to present needs and in an attempt to improve existing designs. It should also demonstrate to them they can be involved in the process and should not accept that any device might not be improved.

BASIC MECHANISMS THAT HELP TO USE ENERGY AND FORCES EFFICIENTLY

Alongside the study of whole machines it is also useful to isolate the different basic mechanisms in order to help the children understand how each works and can be used to make tasks easier. Once each mechanism has been introduced to the children they should be encouraged to find examples in their surroundings and to apply the devices appropriately in their own constructions. The following section reviews significant features of basic mechanisms which can be identified in simple machines and investigated in the classroom.[5]

Wheels, axles and cams

Wheels make it easier to move things by overcoming friction. Ancient cultures found it easier to move blocks of stone using rollers and later invented the wheel and axle which opened the way to much technological development. Most wheels are round and have a central axle. However cams either have the axle off-centre or are oval in shape, producing an up-and-down action as in some pull-along toys in which an animal bobs up and down as the wheels move along the ground.[6] (See Figure 6.1.) *Did children notice the difference?*

Pulleys

Pulleys are wheels with a grooved rim to hold belts or ropes so they do not slip off. They can be used to transfer a turning motion from one pulley to another as in car fan belts, washing machines, vacuum cleaners and sewing machines. Cotton reels have a similar design and can be used to help children understand the idea of transferred energy. Several cotton reels can be nailed to a board so that they are fairly close and can turn freely. By putting elastic bands around two or three of these the children will find by turning one cotton reel the others will also turn. Pulleys, or cotton reels, of the same diameter joined by a belt turn at the same speed. In conveyor belts the band of the pulley is very wide so that it can be used to transfer goods from one place to another.

If one pulley is smaller than another it will turn several times for one turn of the larger pulley. Belt and pulley systems with different-sized pulleys are often used when an electric motor has to drive a machine. When the motor turns a large pulley it can drive a smaller one which will turn faster, thus using the energy supply more effectively. Unfortunately machines cannot create energy or force: they only deliver it in its most useful form. The smaller pulley may go faster but it does not have so much force (or push). Sometimes it is more important to use the available energy to give a strong force, rather than speed. In a washing machine the motor drives a small pulley which turns the

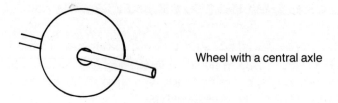

Wheel with a central axle

The non-circular cams make the lorry bob up and down.

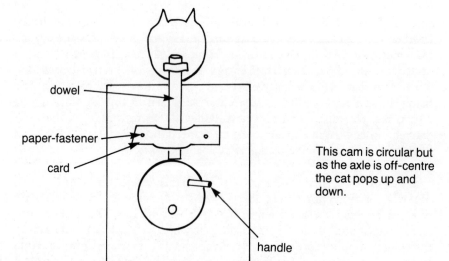

dowel

paper-fastener

card

handle

This cam is circular but as the axle is off-centre the cat pops up and down.

Figure 6.1 Wheels, axles and cams

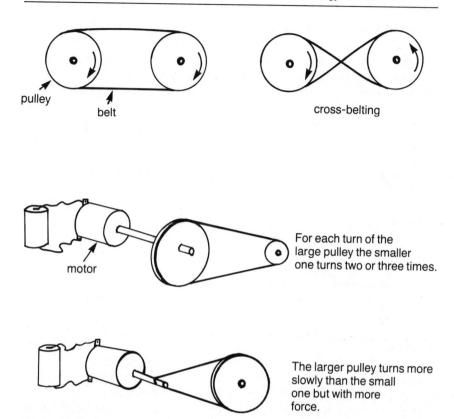

pulley

belt

cross-belting

motor

For each turn of the
large pulley the smaller
one turns two or three times.

The larger pulley turns more
slowly than the small
one but with more
force.

Figure 6.2 Pulleys

large drum which turns more slowly but with greater force so that it can move the heavy water and clothes. (See Figure 6.2.)

Pulleys can also be used to change the direction of movement. If the children join their cotton reels with a twisted band they will find the reels will move in opposite directions. For example this might be used in a rubbish sweeper where the wheels need to turn forward while the spinning brooms are required to turn in the opposite direction to direct the rubbish backwards into a collecting bag. Cross-belting also helps to reduce the amount of slip because more of the belt surface is in contact with the pulleys.

Pulleys and ropes make it easier to lift heavy loads by changing an upward lift into a downward pull. A simple pulley system could be set up by using netball posts or a convenient roof beam. The children will find it is much easier to lift the bucket using the pulley system than by themselves. With some prompting the children should realise that when they try to lift the bucket by themselves they are having to lift against the force of gravity, whereas when they pull down on the rope which goes through the pulley they are using gravity to help them. Two or more pulleys could be set up for the children to discover that it is now even easier to lift the bucket. This is another occasion where the children may feel the pulleys are creating a force or energy for nothing. However they will find they they have to pull a lot of rope to lift the bucket a short way. If four pulleys are used, the children should find to raise the bucket one metre they have to pull the rope a distance of four metres because each piece of rope between the four pulleys has to be shortened by one metre.

The oldest use of a pulley was probably to raise water from a well and the children could make a model with a piece of dowelling and small bucket. They could also be given problems requiring loads to be lifted or moved, which would involve them in designing and making model cranes, fork-lift trucks and conveyor belts.

Gears

There will always be some slight slippage of a belt between two pulleys so that available energy or force is lost. In some cases this can be a safety feature as the motor will still work and be undamaged if the mechanism jams. In other situations slippage is undesirable. This can be overcome by having a toothed wheel and a chain, as in a bicycle. The toothed wheel is often called a cog but its correct name is a gear. (Cogs are the teeth around the gear.) Gears, like pulleys, can transfer movement and forces; increase or decrease forces; and speed up, slow down and/or change the direction of movement.

The children can make a set of moving gear wheels out of serrated bottle tops by hammering the tops onto a piece of wood so that as one turns another will move. It is very important to place the nail in the centre of the top and it is quite difficult to position more than four tops so that they all move. If the children look at the relative movement of the tops they should find that if one is turning anti-clockwise the adjoining one will turn clockwise. Some constructional kits have gears of different sizes that can be investigated. If gears of different sizes are used and marked in some way the children will find that they only need to turn the large wheel once to make the small one turn several times. This can also be seen on a bicycle where the pedal turns a large wheel that drives the chain which turns a small wheel. For each rotation of the large pedal this small wheel turns several times. The small wheel is fixed to the rear wheel so as the cyclist turns the pedals the machine increases the speed the back wheel turns. (See Figure 6.3.) As was discussed in the previous section, a large force (pressure on the pedals) on a big gear can be used to drive a gear quickly but the force is less.

There are special gears that change the direction of movement. Simple hand drills have gears to change a horizontal movement to a vertical one. By examining a hand whisk the children should be able to see both how direction of the handle's turning movement is changed and how every time the handle

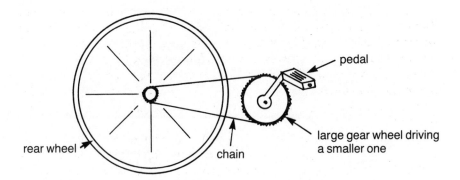

Figure 6.3 Gear wheels on a bicycle

is turned once the whisks turn several times. The beaters of an egg whisk need to turn fast but do not need a lot of force. Again it is not possible to have something for nothing. Mechanisms that increase either a force or speed involve decreasing the other.

Wedges and screws

Another way of lifting a weight more easily is to use a slope. If the children are given a heavy bag of sand and asked to try whether it is easier to lift it directly onto a table or to drag it up a plank of wood, they should appreciate it is much easier to use a slope. Less force is required to lift the bag but it does have to be moved over a greater distance. The huge stone blocks used in the Egyptian pyramids were probably pulled into place up sloping ramps.

A wedge is a special type of slope. The sharp edges of a wedge are formed by two inclined slopes or sides. It is easier to drive an axe into a block of wood by first making a thin narrow cut which then gets increasingly wider. (See Figure 6.4.)

A screw is rather like a small winding slope which also enables a large force to be applied steadily. In the same way, a winding road provides a very gentle slope up a mountain so is easier to climb than the direct route, although of course there is farther to walk. It is easier to get an airtight fit with a bottle top which has a screw-thread than by just pushing one on. The screw can also be found in vices, G-clamps, mincers and inside taps, all of which provide a large controlled force. One of the earliest uses of the screw was invented by Archimedes to raise water. One end of the screw device was put in the water at an angle and turned, which had the effect of lifting up the water. (See Figure 6.4.) The children may also notice that all screws change a turning movement into a linear one.[7]

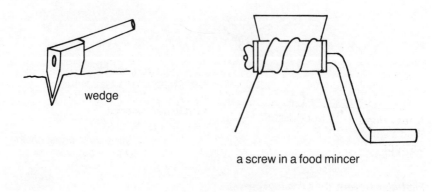

wedge

a screw in a food mincer

Archimedes' screw

Figure 6.4 Wedges and screws

Levers

When the lid of a tin is difficult to remove it often helps to place a spoon handle under the lip of the lid to lever it open. By pushing down on one end of the spoon the lid is pushed upwards by the other. Here the lever changes a downward movement to an upward lift. A lever is a rigid rod which is balanced on an pivot called a fulcrum. In this case the hand is providing both the downward push and the pivot.

Children can be challenged to find a way of lifting a brick off the ground with only one finger, using a piece of dowel and a plank of wood. This can be done if the brick is placed at one end on the plank, the dowel is placed beneath the plank near the brick, and the finger pushes down, as far away as possible, at the opposite end of the plank. The plank is the lever with the dowelling acting as a fulcrum. By looking at the action of the lever the children should see that the heavy weight is lifted only a short distance but they have to push down over a far greater distance, although this needs little strength. (See Figure 6.5.) Using a lever, as with other machines, does not create energy or force, it places the effort applied at the most practical place and in the most convenient direction.

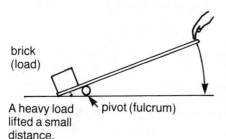

brick
(load)

A heavy load
lifted a small
distance.

pivot (fulcrum)

By pressing down at the end
of the plank little effort is
needed, but the distance the
plank has to move is greater
than under the brick.

First-class levers: the fulcrum is central.

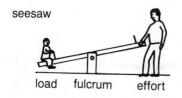

seesaw

load   fulcrum   effort

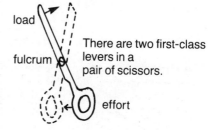

load

fulcrum

effort

There are two first-class
levers in a
pair of scissors.

Third-class levers: the effort is central.

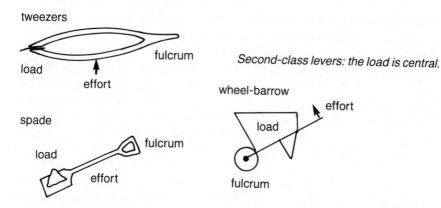

tweezers

load

fulcrum

effort

Second-class levers: the load is central.

wheel-barrow

effort

load

fulcrum

spade

load

fulcrum

effort

Figure 6.5 Levers

There are three types of lever. A seesaw is a type of lever where the centre is the fulcrum. The child who is pushing up at one end is providing the energy or effort and the child at the other end is the load that is being lifted. Not all fulcra are in the centre of a lever. For example a wheel barrow has the fulcrum at one end where the wheel acts as the pivot, with the load in the centre. Tweezers and spades are yet other types of lever. When the gardener is lifting a load of earth with a spade, the handle is acting as a fulcrum and the effort is in the centre of the spade.[8] (See Figure 6.5.)

Linkages join several levers together. They can place an effort in a more convenient location and enable one action to be repeated. When the pedal in a pedal-bin is pressed, a horizontal arm under the bucket pushes up a vertical rod at the back of the bin which lifts the lid. Other linkages are found in umbrellas. As the slider is moved up the shaft, the linkage forces the umbrella open. Similar linkages used to open and collapse devices include rotary washing lines and collapsible music stands. The covering devices on a flute and hammers inside a piano are controlled by linked levers, inaccessible windows can be opened by similar devices, and linked levers are used in grabs for picking up items out of reach. Parallel linkages, as on the side of the shelves of a tool box, act together to enable an action to be repeated.[9]

CONTROLLING MACHINES

All aspects of technology concerned with the release and use of energy are controlled in some way, from the way a lever is used to extremely complex devices. By studying control in simple machines children may be able to apply the same concepts to more complex systems and situations in later years.

Once a machine or system has been designed and connected to a source of power, its control is carried out in one of two different ways: manually when humans are needed to manipulate the parts in some way; or automatically where the system keeps on running until the purpose of the action has been achieved. The children could look for examples of existing machines and how their use is controlled. For example they themselves provide the control for string puppets and bicycles. Mains water taps, kitchen taps and mixer taps in showers all enable water flow to be controlled manually. Switches are another common method of control such as on microwave cookers, gas ovens, electric lights and video recorders. As part of electrical work in science the children will undoubtedly make and use switches, using either commercial or junk materials, which can be incorporated into technological projects. The children might make a torch, theatre lights, or car with headlights which have switches for turning their device on or off.

Slightly more complex control systems can be set up so that they are triggered by some outside phenomenon such as a burglar alarm which sounds when a door is opened. One solution is to have a strip of metal foil attached to the top of the door close to the hinge which, as the door opens, touches another metal strip hanging down just behind the door thus completing a circuit and setting off an alarm. Another project could be to make a flood alarm system which might consist of a wire fixed to a ball covered in foil which, as the water level rises, touches a second wire leading to the alarm. The children might also try making a switch that is operated by a strong wind or one that is triggered by a trip-wire.

A light that flashes at regular intervals (for a model ambulance or lighthouse) without needing to be constantly switched on and off can be made by glueing a line of foil onto part of a cotton reel. A second strip of foil joined to a circuit with a bulb should be placed so that it touches the cotton reel. The cotton reel is turned by the movement of a model or motor and, when the foil touches the second strip, the light will come on but will go out as the uncovered part of the reel comes into contact with it.[10]

Feedback

In order to control most machines there needs to be some sort of observation and feedback. Someone using a lever to remove the lid of a tin will watch and adjust the amount and direction of pressure exerted, depending on how quickly and easily the lid comes off. In the same way a person will consider the temperature and amount of light in a room before deciding to switch on or off the light or heater. This decision-making process is expressed in a diagrammatic form in Figure 6.6.

If a human being is part of the process the whole procedure is usually very complicated as often more than one input and feedback path is controlled. In the bicycle the rider can vary the steering, gears, pressure on the pedals and body position after observing the road conditions. (See Figure 6.7.) Other examples where people are part of feedback systems include adjustments of volume, station and tone to a radio; of volume, brightness and colour on a TV set; or of the hot and cold water mix in a shower.

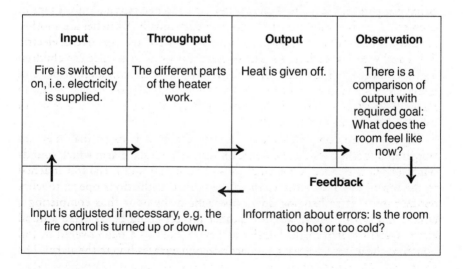

Input	Throughput	Output	Observation
Fire is switched on, i.e. electricity is supplied.	The different parts of the heater work.	Heat is given off.	There is a comparison of output with required goal: What does the room feel like now?

		Feedback	
Input is adjusted if necessary, e.g. the fire control is turned up or down.		Information about errors: Is the room too hot or too cold?	

Figure 6.6 Controlling a heater

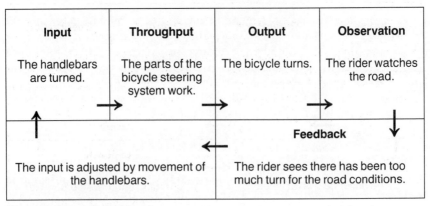

Figure 6.7 Steering a bicycle

Automatic control

Automatic control involves a human deciding what the desired output should be and setting up the device in the first place, but then the system handles the observations and adjusts the input accordingly. This does not necessarily have to be handled by computer. Since water is constantly flowing into the house to the storage tank it is important to have a method of control that does not need people to be present all the time. This is done by the ballcock and the overflow pipe. When a lot of water is used the ballcock falls and opens a valve which lets the water in. As the water rises, the ballcock rises with it and has the effect of closing the pipe.

Many automatic controls are, however, linked to computers. The automatic control system on a modern boiler has a sensory device which registers when the temperature falls below a particular temperature and switches the boiler on. When the desired temperature is reached it switches the boiler off. (See Figure 6.8.)

Children can look for other examples of automatic control around the school and house, such as the oven which turns down when the correct temperature is reached, and the washing machine which turns off the water supply when the tub is full. Some traffic lights have a vehicle detector on the ground on roads approaching a junction which counts the cars, and this information is processed by the computer which controls the lights. Supermarket automatic doors use a pressure pad or infra-red sensor. Cows entering a milking parlour sometimes wear tags which are scanned as they enter so that each cow is automatically fed the correct amount and type of food. Temperature, light and humidity are often automatically controlled in commercial greenhouses.

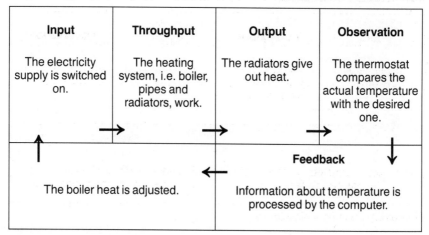

Input	**Throughput**	**Output**	**Observation**
The electricity supply is switched on.	The heating system, i.e. boiler, pipes and radiators, work.	The radiators give out heat.	The thermostat compares the actual temperature with the desired one.

Feedback

The boiler heat is adjusted.	Information about temperature is processed by the computer.

Figure 6.8 Automatic system for a boiler

Closed- and open-loop systems

Most of the control techniques discussed so far are examples of closed-loop control systems, as they depend on either human or mechanical observation of the environment and a subsequent readjustment. On the other hand open-loop control systems do not rely on any feedback and are independent of the output. For example, where traffic lights are working on a time switch and are unaffected by the cars on the road, an open-loop is used. This design depends on the timing having been pre-set sensibly, and may involve prior surveys and investigations.

First experience of computer control can be provided in the early years by using programmable toys and robots. When the application of computer control is explored in greater depth these experiences can be recalled. The way the classroom computer controls the monitor, printer and floor turtle can also be pointed out. Further practical opportunities for setting up their own control systems can be provided by giving the children the opportunity to link the classroom computer to models they have made. They may well find that the computer does a task more easily than a manual method; for example children sometimes find using a control box to get a light to flash on and off much easier than making a mechanical device.[11]

When designing a new machine it is not only important for the designer to understand how mechanical devices can be linked and controlled to create an effective system or product, it must also be user-friendly. An exciting new machine which may appear to be a highly desirable product, such as a video recorder, is of little value if it is too complicated for the user. Therefore as children become more competent with using and controlling appropriate mechanisms in their designs they must also be encouraged to think about how they will be used and by whom, and to adjust the design accordingly.

NOTES AND REFERENCES

1 Jarvis, T. (1991) Chapter 4 Discovering energy and forces in play *Children and Primary Science* London: Cassell.
2 West, J. (1990) 'Turn back the clock' *Junior Education* January, pp. 34–5.
3 al-Hassan, A. and Hill, D. (1986) *Islamic Technology: An Illustrated History* Cambridge: Cambridge University Press.
4 Lego Technic 1 Activity Centre and Teacher's Guide has many open-ended problems requiring the application of simple mechanisms.
5 Bindon, A. and Cole, P. (1991) *Teaching Design and Technology in the Primary Classroom* Glasgow: Blackie; and Jarvis, T. (1991) Chapter 12 Models and machines *Children and Primary Science* London: Cassell contain useful sections on making and understanding different mechanisms.
6 Johnsey, R. (1990) Chapter 4 Wheels and levers *Design and Technology Through Problem Solving* Hemel Hempstead: Simon & Schuster includes ideas for children to make models using cams or eccentric wheels.
7 Dunn, A. (1991) *Simple Slopes* Hove, East Sussex: Wayland has well-illustrated examples of how wedges and screws are used.
8 Ollerenshaw, C. and Triggs, P. (1991) *Levers* London: A. & C. Black gives simple activities to help children understand levers. Dunn, A. (1991) *Lifting by Levers* Hove, East Sussex: Wayland has well-illustrated examples of different types of levers. Both books are suitable for Key Stage 2 children.
9 Oza, V. and Chandler, M. (1990) *Starting Design and Technology: Mechanisms* London: Cassell gives good resource information and ideas on machines and linkages in particular.
10 Ideas for developing ideas about manual control can be found in Fennell, M. (1989) *Starting Design and Technology: Energy and Control* London: Cassell: and Johnsey, R. (1990) Chapter 6 Electrical switches *Design and Technology Through Problem Solving* Hemel Hempstead: Simon & Schuster.
11 Straker, A. (1989) *Children Using Computers* Oxford: Blackwell. NCET, Sir William Lyons Road, Science Park, University of Warwick, Coventry CV4 7E2 also produces excellent material on control technology.

FURTHER READING

Macaulay, D. (1988) *The Way Things Work* London: Dorling Kindersley.
McCloy, D. (1984) *Technology Made Simple* London: Heinemann.
Technology for Teachers Course Team (1975) *Fundamental Concepts in Technology II* Milton Keynes: Open University Press.

Chapter 7

Structures

Young children need to learn to recognise and make models of simple structures and know that objects are changed by forces being applied to them. As the children mature they should increasingly be able to identify pattern in structures and should make and test two- and three-dimensional models for themselves. Much of this work can be combined with scientific work on forces and investigations on everyday constructional materials. Children need to examine and understand how forces affect existing structures by identifying internal and external forces on various constructions; investigating how different materials react when a force is applied; and examining how shape and method of construction can increase the strength of a structure. Once children have a grasp of some of these concepts they should be able to apply this knowledge in their own constructions.

A structure provides support by holding something up; reaching across or spanning a gap; and/or containing or protecting something. The girders of an oil platform hold up the oil rig, a lamp post carries a light and walls support a roof. A crane arm, bridge, roof and wings of an aeroplane all reach across or span a distance. Cans of drink, milk crates, plastic bottles, egg boxes and car bodies are all structures that contain items. The children could try to find examples of different structures in books and in their environment and try to identify some of the forces acting on them.

Each structure must be strong enough to resist internal and external forces.[1] To keep the shape of the structure these forces must balance. If the force on the outside of a drinks can is greater than the outward push of the structure and the liquid it will collapse. Some of these forces are stationary or static and others are dynamic. Static forces are inherent in the actual structure and in any steady, consistent external forces such as the load of a roof on the walls. By examining the loads on a bed the children can identify these different types of forces. The material that makes the frame and mattress will be pulled down by gravity. The children can see this happen if they roll some soft plasticine into a long snake shape and place it across a span between two blocks. It will sag in the middle showing how objects have body weight and bend under their own weight. The design of the bed must be able to resist this. In addition the

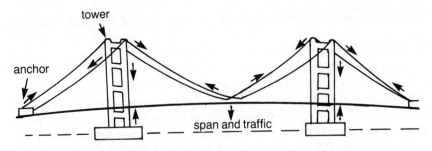

Figure 7.1 Forces in a suspension bridge

bed has to take the fairly still and steady load of a sleeping person. However the structure of the bed will probably also have to withstand the dynamic forces of the child bouncing up and down.

These different forces can also be identified in the construction of suspension bridges. The pull down of the bridge's span, through gravity, must be balanced by cables that pull upwards. If some of these cables pull too much, the structure will not keep its shape. These cables pull the towers inwards and their effect is counteracted by the cables pulling outwards to the anchors. The towers are pushed down by the weight and action of the cables which is balanced by the force of the earth pushing upwards against the base of the towers. (See Figure 7.1.) Unless these forces are balanced there will be movement and the bridge will collapse. Symmetry plays an important part in the overall strength of any construction as it is easier to balance all the component forces when the structure is symmetrical. In addition the bridge also has to withstand moving or dynamic forces such as the weight of people and traffic going over the bridge and the action of the wind and waves. The children should be able to find other examples of dynamic forces on structures such as walking on floorboards, wind and rain against walls and windows, trains over railway tracks, and fairground rides.

Although safety is likely to be the paramount design factor, as people must be able to use the product without the danger of its breaking or collapsing, the designer also needs to think about cost, appearance, and indeed whether it is possible to construct it at all. A bridge design may appear perfect but if the pieces cannot be assembled across a span it is useless. Children can consider the latter problem by being given the challenge of getting the first line across an imaginary steep gorge in order to build a suspension bridge. They might suggest ideas, such as using a catapult, model aeroplane or kite, some of which can be made and tested.

MATERIALS

Architects need to minimise the amount of material used in a building so that the cost is not too high but they must also ensure high standards of safety. Consequently it is important to test materials that may be used in a construction to assess how strong and rigid they are. Rigidity is the ability of a structure to withstand being bent, twisted, stretched or changed in shape. Strength is how much force a material can stand before it breaks. Different materials act in different ways when bent or compressed. The children can try bending and squeezing different things, such as chalk, a pencil, thick wire and rubber. They should find that the chalk will snap suddenly but the pencil will bend initially then splinter and break. The wire will bend but will not return to its original position. The rubber bends and returns to its original shape but it may also tear and break.

If materials are weak in tension (stretched) and/or under compression (squashed) they will break under heavy loads. To help the children to understand the difference between tension and compression they can be asked to press their hands together to feel the sensation of compression and then pull outwards to experience tension. If they also squeeze a sponge they will see that compression makes things get shorter and fatter whereas tension makes them longer and thinner. If the children bend a rubber they will see that when a material is bent the outer part of the bend will be under tension whereas the inside is being compressed.

Metal and wood are strong in both compression and tension but are liable to rot or rust. Natural stone and concrete are very durable and are strong in compression but are weak in tension. When describing and evaluating different structures the children should try to suggest why the designer or architect chose the material and should consider the properties of different materials when making suggestions for improvements and when planning their own designs.

Wood

Balsa wood is very good for tests in the classroom as the children can cut pieces relatively quickly and easily, although it must be remembered that it has very different characteristics to most wood they will meet. If wood is to be used in a roof or bridge it is important to know how much weight, e.g. tiles or load of traffic, it can carry before breaking. The class could be asked to discover the minimum amount of wood needed to make a model bridge across a gap between two tables to carry 500 gm, 1 kg and 2 kg, or different groups of children could test the breaking point of different lengths, widths or thicknesses of wood and report back to the class. One method of testing is to place a strip of wood across the gap between two tables. A bucket is hung on the middle of the strip and the ends are clamped to the tables. Weights are slowly added to the bucket until the wood breaks and the various breaking points can then be recorded and compared. The children should wear goggles to protect their eyes in case the wood jerks upwards and splinters as it breaks. Newspaper placed beneath the bucket will protect the floor.

If the children test the breaking points of wood other than balsa wood, they should notice that it bends and can return to its original shape before the breaking point is reached. Where structures, such as boats or tennis rackets, have to be flexible, this property is an asset but other structures need to be firmer.

Concrete

Concrete is a kind of artificial rock made by cementing together gravel or crushed stone (aggregates) and sand. Unlike wood, wet concrete can be poured into moulds of almost any shape and size. Because it can be made into whatever shape is needed it is far more useful than natural rock, which is difficult to extract from the ground and afterwards has to be worked to the required shape. The children can easily make different mixtures of concrete and pour them into moulds such as yogurt pots or specially made wooden ones to make narrow beams which can be tested to breaking point. After about half an hour the concrete starts to harden, in a few hours it is hard to touch and it continues to harden for some weeks. Indeed after about three or four days it will be extremely difficult to break concrete beams made in the classroom at all.

The strength of concrete is produced by chemical reaction between the cement and water. This proportion is very significant as too much water makes the concrete weak. However water cannot be reduced too much, as it is not only required to react with cement but it is also needed to make the concrete flow easily so that pockets of air can be forced out. Air holes will provide places of physical weakness where the concrete might break. It is possible to add a plastic which helps the particles slide over each other so that concrete can be pumped into very narrow apertures without using too much water. Unfortunately in this case the concrete takes longer to harden and take heavy weights. The more cement that is added the stronger the mixture, but as cement is the most expensive ingredient the manufacturer will wish to use the minimum possible which still ensures a good-quality product. Consequently when making concrete the proportions of the various ingredients are varied according to the strength required for each individual situation. It can be seen that producing concrete of a high quality at a reasonable price is a highly skilled business and that testing and product control need to be important aspects of the industry.

The forces of compression and tension are particularly significant in the use of concrete, as it is a very strong material under compression but weak in tension. By using another material with different properties it is possible to overcome concrete's weakness in tension. In reinforced concrete, steel rods are imbedded in the mixture. These bars are very strong under tension so will resist if the beam is bent. Unfortunately, unlike concrete, steel rusts in air and loses its strength; therefore steel bars used in reinforced concrete are placed below the surface of the concrete where this should not happen. The children can make their own reinforced concrete by adding florists' wire to some of their concrete bars and comparing them with bars made of the same mixture without wire. By combining materials with different properties, builders can exploit the advantageous properties of the materials and minimise their limitations.

The children might discuss when they could use plain concrete and when it would be better to use reinforced concrete. The former is suitable for retaining walls, dams and other structures that rely on their great mass for stability. Foundations and modern motorways are other good situations as they are not going to be bent. However, in beams and bridges where the concrete is under compression and tension, reinforced concrete is preferable.[2]

STRENGTHENING STRUCTURES

Strengthening materials by folding or bending

Thick solid material is stronger than a thin strip of the same material, but the former may be too heavy for the overall structure. By folding or bending thin material it is possible to increase its strength without additional weight. A horizontal piece of card will resist bending if its opposite sides are pushed together horizontally, but once the card starts to bend it will buckle upwards or downwards. If the card is folded at right angles along its length the resultant piece can no longer bend easily up or down and therefore has increased strength in all directions. This principle is just as applicable whether the structure is made of steel, concrete or paper. By investigating paper structures the children can discover how materials can be strengthened by bending or folding them.

The children might investigate ways of using one or two pieces of A4 paper to make a bridge to span a gap of 20 cm and carry a weight of 100 gm. They should discover that the flat piece of paper will carry very little weight, but by folding the paper into a concertina or box shapes, or by cutting the paper into two or three long strips and making tubes, the strength is increased considerably. These shapes provide stronger structures than a flat piece of paper as they resist twisting and bending in any direction. Bundles of tubes are even stronger. The children can be given 20 straws and asked to tape them to make a flat bridge and then test what weight the bridge will take before sagging. Another 20 tubes can be grouped into two or three bundles tied with elastic bands and tested in the same way. The children will find these have even greater strength.

There are many examples of these methods of strengthening of materials in the school and environs which the children can try to identify. It is common to see steel girders folded into right angles. Bicycle frames use tubing to make the machine light but strong. Plastic chairs often have their edges curved over and hollow tubular legs; plastic buckets and tidy boxes frequently have the rim or edges thickened, turned over or folded into a girder shape. Thin plastic food and drinks cartons use systems of folds or curves in the material to give strength.

Investigating rigid shapes

Materials can also be increased in strength and rigidity by the way they are joined. For example an open four-sided toothpaste box will collapse when pushed down on one corner but a triangular chocolate box, also made of card and of a similar size, will resist collapse. The children can explore this idea further by joining strips of card held together with paper fasteners or lengths of plastic rods from a construction kit to make a variety of simple two-dimensional shapes, including squares, rectangles, triangles and pentagons. If they then push their shapes gently on one side they should discover that a framework made of triangles cannot be pulled or pushed out of shape. A variety of three-dimensional shapes could also be made with straws fixed with short pipe cleaners or folded strips of card to discover that the triangle is as important in these constructions. The children should find that triangular shapes providing strength are found in a variety of places such as in the spokes and frame of a bicycle, in the design of pylons, playground equipment, gates and tents, and in roof supports. Many beam bridges are strengthened by a triangular framework at the sides or beneath the bridge.

An arch is another method used to stop the downward bend of a flat beam. The children can discover this by placing a flat card between two bricks and comparing the weight it can carry with a similar one supported by an arch. The weight of the load is spread by the arch to the two side bricks. However, if light blocks are used instead of bricks, they will be pushed apart showing the need for very solid supports. Domes, as in roofs and crash helmets, are analogous to many arches around a central point so are strong and use material economically. Even with fragile material like egg shell this structure gives great strength, which can be demonstrated by collecting egg shells and cutting them to make four halves which are then placed arch upwards with a piece of hardboard on top. The children will see that this structure can hold several books before the eggs finally break.

Method of construction

The way the components are put together is also significant. This can be explored by looking at how bricks and concrete blocks are fixed together to discover that overlapping increases the strength of the whole structure. The children could design and carry out a test to compare the strength of a wall made of overlapping toy bricks with one made of bricks that do not overlap. To ensure the test is fair it is important to apply the same force to each wall in the same way. The children may decide to set up a device that swings or rolls increasing weights against the walls and then count how many bricks have been dislodged. The common brick patterns or bonds could be compared in the same way.

EVALUATING AND MAKING DIFFERENT STRUCTURES

Studying the design of bridges is a valuable way of exploring structures in greater depth. The children might evaluate actual bridges both in the past and in the present, including stones with slabs placed on top (clapper bridges), Roman arched bridges, rope bridges, wooden trestle bridges, concrete motorway beam bridges, suspension bridges and cantilever bridges. They could also include notable disasters and why they happened, such as the Tay Bridge which failed due to faulty casting and poor design. The children should also be encouraged to consider the aesthetics of different designs, perhaps by producing silhouettes on simple backgrounds, string prints, or stylised reproductions in fabric and stitching so that they can focus on the line and shape.

Finally the class might be given an open-ended task combining the different elements of bridge building by being asked to build a bridge to span a 60-cm gap between tables within a limited time. Materials could be purchased with some sort of token with a limit of ten per bridge at a fixed prices for different materials, such as one token for four sheets of newspaper, one piece of card, two sheets of A4 paper, 30-cm of sticky tape or string, etc. The completed bridges can then be judged on strength, beauty and economy.

Packaging has the advantage that examples are easy to collect, handle and test. The children can also make their own to satisfy specific requirements such as a container to carry a fragile chocolate egg or soft fruit. The project might start by carrying out a survey to examine different packages to identify how the purpose of the container affects the materials used and its shape. Questioning will help to identify some of the significant factors: Does the container have to carry liquids or solids? Will it need to be heated or frozen? Are the contents fragile? How does the shape of the contents affect the package? Does the container need to be reused, as in glass milk bottles, or recycled? What forces does it have to withstand and what means are used to strengthen the material? Does it need to be stacked? Has the material been used economically? By exploring these questions and other similar ones the children should be able to evaluate existing containers and suggest improvements which should help to ensure that their own designs are more carefully thought out.

When studying the design of buildings, different types of transport, play equipment and furniture, concepts relating to their structure can usefully be included. In each case the children should consider the properties of the material used for construction and the way it is shaped and constructed to withstand both static and dynamic forces, such as cars, ships and aeroplanes having to withstand the fairly static forces of passengers and freight and the dynamic forces of varying weather conditions and the possibility of collision. By describing and evaluating a variety of manufactured structures the children should be able to take these factors increasingly into account in their own creations.

NOTES AND REFERENCES

1 National Curriculum Council (1992) *Knowledge and Understanding of Science: Forces: A Guide for Teachers* York: NCC is a very useful book designed to help teachers to increase their knowledge and understanding of science ideas relating to forces and includes a section on the forces in structures.
2 Higgins, R. (1972) *Materials for the Engineering Technician* London: English Universities Press.

FURTHER READING

Bull, R. (1989) *Starting Design and Technology: Structures* London: Cassell.
Jarvis, T. (1991) *Children and Primary Science* London: Cassell.
Kincaid, D. and Coles, P. (1979) *Science in a Topic; Roads, Bridges and Tunnels* Amersham, Bucks.: Hulton Educational.

Chapter 8

Examining economic enterprise

Throughout technology there are considerable overlaps with the cross-curricular theme of Education for Economic and Industrial Understanding as both subjects can help children develop an interest and awareness in economic and industrial affairs, and to understand the key concepts of production, distribution, supply and demand. Young children can learn that goods are made, distributed, sold, advertised and bought. They should be able to appreciate that the creation of products is limited by costs of materials, equipment, available people, their skills and time. Study of industries also enables them to see design and making in 'real life' and provides many contexts for their own projects.

Such a study does not necessarily mean accepting that economic values are the most important in a society. On the contrary, it enables children to discuss related issues in an informed and balanced way. Children who are introduced to the workings of factories, farms, laboratories, offices, shops and transport organisations while they are comparatively young are less likely to be either complacent or overawed by them when they enter the adult world, and as a result are more able to adopt or adapt them to their own and society's advantage'.[1] Children will become more aware of the need to carefully examine and evaluate manufactured products rather than be over-influenced by persuasive advertising. Hopefully they will also develop a concern for scarce resources and a sensitivity regarding the effect of economic choices on the environment. In addition, by giving children a number of varied experiences of enterprises, they are more able to make informed choices about using their talents and qualities in their future careers. It may also be possible to use the experiences to counter stereotypical expectations related to gender, race or class.

Primary schools have for some considerable time involved members of the local community in their work with young children through the popular topics of 'People Who Help Us' and 'Our Town'. These topics can be extended to enable young children to examine retail outlets and other familiar businesses in the community. The school could then branch out to visit and investigate a variety of industries and invite people from different business concerns to school. The children can also have opportunities to simulate or even set up their own enterprises. Teachers should find that the primary system is in many ways more in tune with business approaches than many secondary schools, which should help them to set up successful simulations and mini-enterprises. This was seen when the Fulmer Industry Project first started working with primary schools, when they commented that many primary schools operated in a similar climate as the engineering design office, as they worked in groups on open-ended problems, on projects related to the world outside school, crossing subject boundaries and letting interest, inquisitiveness and enthusiasm motivate them.[2]

Some teachers and parents are concerned that industries are unsafe environments for young children, but by choosing from a huge range of possible enterprises many suitable establishments can be found. A small work place employing three or four people can be just as valuable as one that employs hundreds. However the industry should preferably be one providing a service or producing something the children relate to, such as bicycles, toys or domestic items. Complex processes which are very difficult for very young children to understand should be avoided. Successful projects have included studying local stores, fish and chip shops, hairdressers, garden centres, fast-food outlets, banks, post offices, hospitals, travel agencies, garages, hospitals, farms and hotels.

Past projects indicate that even very young children can get a good understanding of how a company works, the role and functions people perform, how wealth is created by industry and the advantages of working in a team. Forsbrook Infants School, for example, after having visited a manufacturing establishment successfully set up their own mini-enterprise to design and sell a product.[3] Indeed the children often surpass the expectations of industrialists and many teachers by their sophisticated questioning and understanding.

CHILDREN'S UNDERSTANDING OF ECONOMIC CONCEPTS

Young children are not strangers to industry and many may even have greater knowledge than their teacher about the enterprise their parents are involved in. Children visit shops with their parents and watch the transactions. They see industrial products around them, and are familiar with and are influenced by advertising. They are aware of parents going to work or wanting to have work and know it is an activity that most adults rate highly for which they get money. However children's perceptions are gathered piecemeal, often from secondary sources like television and overheard conversations, leading to incomplete or incorrect ideas. For example, before investigating economic enterprises, children often have very erroneous stereotypical views of industry. They often equate 'industry' with 'manufacturing' and think of it as being noisy, dirty and dangerous, which makes girls in particular apprehensive about working in 'industry'. Activities investigating different industries should enable the children to develop a more balanced view in which manufacturing industry is seen to be only one part of industrial activity in Britain, there being a wide range of others including the service, construction, retail, distributive, agricultural and public sector industries. They should also discover that many manufacturing processes are clean and quiet.

Research being carried out in different countries on how children's understanding of various economic concepts develops can be used by teachers to choose the most appropriate activities and issues to explore with the children. Children's reasoning about ideas such as ownership, profit and price tends to follow a developmental sequence and shows a gradual improvement with age, although children of the same age may be at different stages, as their development can be influenced by their competence in other subjects such as mathematics. Those concepts that relate to concrete examples or are in the children's experience develop more quickly, indicating the importance of providing practical first-hand experiences. Profit in particular appears to be a difficult concept for children. In order to get a good grasp of all these ideas children need to build up their experiences over several years and in different situations.

By knowing some of the children's common misconceptions or alternative views the teacher is better able to provide experiences and ask questions that extend and develop the children's comprehension more rapidly. Consequently it is intended to review some of the research on children's economic concepts so that teachers can ascertain and effectively develop their children's understanding.

Work

Children tend to have a limited concept of 'work'. Their views could be determined by asking them to list what they identify as work at school and at home. Ideas about differences between work and play could also be explored. The children might suggest work is helping with household chores, shopping, practising a musical instrument and some school work (PE, art and practical activities in science are frequently excluded) and is often described as boring, compulsory, physically hard or mentally difficult, must be completed, and not as much fun or as messy as play. As children tend to link the idea of work with physical effort they often find it difficult to appreciate that people in offices or even teachers are working! It is important that children see that this is a simplistic view. They might be asked whether they think the professional footballer in a league game, the swimming-pool lifeguard watching on the pool side, or the manager sitting and talking in a meeting are working. Such questioning can help the children to appreciate that many adults enjoy elements of their work, work can be intellectual, and plenty of recreational activities are physically and mentally demanding. One way children may try to distinguish adult work from recreation is by whether it is paid. However, although a guide, this is inadequate as some adults' work is unpaid, such as housework, child care and voluntary work. A broad view will only develop over time through class discussions and visits to and from representatives from different enterprises, such as office workers, gardeners, artists, people from a factory floor and those who deliver meals to old people.

While most children realise most adult work is paid they are very confused about the source of the money. Children build up concepts by making generalisations from their own experience and they rarely see adults being paid. Therefore they tend to assume that when money is exchanged it could be used for wages. They may think that teachers are paid from dinner money or may even do other jobs in the evening for money. It might be worth letting children become more aware of teachers receiving salary slips and how these indicate money being paid into a bank account.

Exchange and profit

In the same way children see money being put into a shop till but rarely see it being removed. Therefore, unless it is pointed out to them, they are unlikely to link this money with the shopkeeper's wages or need to buy other goods. Jahoda[4] and Linton[5] investigated the idea of exchange and profit in the context of shop transactions with young children and suggest there are several stages of reasoning.

They found that although children around four or five years old associate money with buying and most know that non-payment is illegal, they do not understand its proper function. Young children commonly think that the transaction is some kind of ritual with the idea that money has to be given so the shopkeeper can give some back. In their perception, once money is put in the till it stays there or might be given to charity. If children are asked how goods get into the school, they talk about factories, lorries or even the government but do not see that money has got anything to do with the process. One way the teacher might assist understanding is by developing play in the class shop to encourage the children to imagine counting up the cash and taking it to a bank at the end of the day, and suggesting the 'shopkeepers' pretend to go to a wholesale market to buy the goods for the shop. The children could talk about how they use their money and what happens to it if they buy goods such as sweets or comics. They might also talk about when they use money outside a shop, perhaps for buying a bus ticket, to put in a charity box or to buy a ticket to swim, and discuss what this money is for and what happens to it. When the children visit a shop they should try to find out where the shopkeeper got the goods, who pays for them and where the money comes from. They should be encouraged to ask questions like: What happens to the money in the till? Where does the shopkeeper get money for his/her home or television? How does the person in the shop decide what the items will cost? How do the goods arrive at the store? Who pays the delivery driver?

By early Key Stage 2 many children appear to have the idea of two unconnected systems, often assuming that goods are bought and sold at the same price. The children usually know that the customers pay for goods. They also appreciate that the shop has to buy goods and pay for shop assistants but they do not make the link in order to understand that the money for the new goods and wages comes from the customers. They often suggest it came from the government or jobs. Grasp of the way these two systems intermesh is not usually reached until 10 or 11 years, unless direct questioning triggers off an earlier insight.

Jahoda and Linton found that children from about 10 years old realised that money was taken to the bank and used to pay workers and buy things. The children also showed an increasing awareness that there was a difference between the buying and selling price. However many children, even when they understood the idea of profit, could not see where the owner fitted into the system. When asked where the manager got his money many suggested from another job. It appears that the concept of profit is a difficult one for children and does not occur until about 11 years of age. This seems to be, at least partly, related to the children's strong view of fairness. There appears to be a view throughout the primary years that selling something for more than it was bought for is unfair. Only by 11 are some children prepared to accept that a profit on a sale is good. In this case they endorse the idea that profit should cover expenses of the store and the employees' wages. Anything over and above this is not considered legitimate profit.

Price

When setting up a role-play shop or actually selling goods in school the teacher may find it useful to investigate the children's ideas of how the items should be priced. These views can be used to assess the children's understanding and as a stimulus for discussion. Through such dialogue the children are able to articulate and clarify their ideas, compare them with others and move more quickly towards a greater understanding. Schug[6] and Linton[7] among others have investigated children's ideas about price. Children were asked to name things that cost a lot of money and explain why. They were also asked about things they felt were inexpensive. Another task involved asking which of two items, such as a wrist watch and book; pair of shoes and bar of chocolate, and house and motor car, cost more and why.

The youngest children, about five years in age, used size or heaviness to explain price. They said big things such as houses and cars were expensive whereas small items like sweets were cheap. If an apple was bigger than a peach they expected it to be more expensive but if they were the same size the children felt they would cost the same. Another common explanation for the price was simply because the price tag said so.

By about eight years old, although the children tended to still use physical attributes, they also commented on function and durability. Usually the children focused on only one or perhaps two factors. They felt that useful things and items that worked well would be more expensive. A house, for example, was thought to be more expensive than a car because it took a long time to build and was more important, as people could go on foot if necessary. Refrigerators were said to be expensive because food can be stored in them and they were cold. A bath would be more valuable than a banana as it can be used more than once and lasted a long time, and according to some children apples must cost more than peaches because apples took longer to eat.

By the top of the primary school children suggested that prices were related to production inputs, and the scarcity and popularity of the product. They felt that an object costs more when the materials in it were more valuable, if it had to be transported farther or if it took longer to make. For example the children said a peach would be more expensive than an apple because it cost more money to bring it to Britain; a pencil would be cheap as it was easier to make, whereas shoes would be more expensive because they took longer to make and were complicated to manufacture. At this age children also start to appreciate that an item costs more if it is in short supply or is wanted by large numbers of people.

Supply and demand

Changes of prices for the same item can be very puzzling for children. When asked why prices for the same item fluctuates at different times of the year, very young children are likely to say that it is a trick, to enable poor people to buy things, or to stop customers becoming bored. In order to help the children the teacher might ask how and why things vary in different seasons. They may be able to suggest that the cost of food varies depending on whether people want to eat warm or cold food and on the growth pattern of vegetables and fruit. Other sales are also affected by seasonal change. Tennis rackets and barbecues might be reduced in the winter months as there is little demand for these items then. The children could also discuss why shops have sales and should be able to appreciate that these are often to make room for new stock. As children mature they can identify a greater range of factors such as that a poorly selling item or deteriorating food might be reduced, or even that competition between shops and 'inflation' cause price fluctuations.

Ownership and organisational hierarchies

Most adults describe the arrangement of complex organisations in terms of a hierarchy. However many children do not see them in this way. Indeed they may well be puzzled as to who is in charge or is the owner. Many children, for example, will identify the caretaker or secretary as the senior adult in the school and are highly unlikely to even consider an outside body. Early-years children tend to think the owner is the person who is in contact with the object.[8] For example they say the passenger owns the bus. As children's understanding develops they perceive the person who is in control or who directs the production of the goods or service as the owner. They think therefore that the bus driver must own the bus and that he/she has bought it. Later in their development the children are able to distinguish between the owner and employee and say the boss is the owner. Finally they appreciate that there is a hierarchy of command where the boss might be in an intermediate position between the owner and worker.

Ross[9] studied 7–11-year-old children who had visited different enterprises, including hospitals, banks, a post office, supermarket and engineering works, to see how children explained an imaginary organisation of people with specific jobs such as chairman, technical director, chief engineer, engineer, sales director, sales manager and sales staff. Each child seemed to suggest one of three models, one being the sophisticated branching hierarchy common to capitalist management, the other two being linear models. One linear model put the emphasis on functions, e.g. all the people working in engineering were put in one line, followed by a line of all those in sales regardless of status. The other linear model concentrated on grouping individuals of the same status together such as a line of all directors followed by a line of all supervisors. Younger children usually preferred a linear model and girls tended to give higher value to status. Ross suggests that girls are more acutely aware of the disadvantageous status of women and are therefore more conscious of power relationships, which appears to enable them to understand the status relationship of a hierarchy more quickly.

Ross found that the children's understanding depended on the actual experiences they had had in the different businesses and suggests that children need a variety of practical experiences to build up a good grasp of hierarchical arrangements.

VISITING AND STARTING BUSINESS ENTERPRISES

Practical experiences that provide relevant contexts are essential to enable children to develop a good understanding of economic concepts. Unlike the secondary sector, primary children do not have the opportunity for work experience. However a visit to an industrial concern and practical activities in the classroom, such as simulations and setting up their own businesses, can provide the necessary concrete experience young children need to grasp the rather difficult concepts involved. There is a temptation to cover many ideas to maximise a visit or project, but there is agreement from both teachers and industrialists that it is better to concentrate on one or two concepts at a time.[10]

Visits to industrial concerns

Establishing links: Initially the school needs to establish links with a company. The children could help to draw up and distribute a questionnaire asking parents about their work experiences, which if carefully phrased should enable people currently unemployed or retired to answer positively and perhaps share their past expertise. The governors could be approached, or a walk around the neighbourhood might identify suitable establishments which could be contacted, as links with a neighbourhood industry are advantageous especially if the place is known or some of the people are familiar to the children. Additionally, physical proximity for visits by the children and any return visits by the work force will be easier. It is also usually better for most primary children if the enterprise is relatively small so that they can get a reasonable grasp of processes involved.

Preparation: As many business people will not know what to expect from young children, preparation for the visit is as important for the industry as for the children. Consequently the teachers need to be prepared to explain their aims, time involved and reasons why children need this type of experience when making their preparatory visits.[11] The most successful visits have been found to be focal experiences in an extended study which might last half a term or more.[12]

The visit: It is very difficult for the children to absorb all the important features during one visit so it is sometimes helpful to arrange more than one, particularly if the type of enterprise is very unfamiliar. During the visit the children need to be able to concentrate on the experience, so a very detailed work-sheet or the requirement to take vast quantities of notes may be counter-productive.

Following up or recapping the experience: This could include building models of the work place, showing the production process in diagrammatic form; replicating a simple production line in the classroom; and inviting people from the industry to the school. The latter can be particularly beneficial when the children know sufficient about the establishment to ask valuable questions. The visit might also be recalled by the use of a video which could incorporate sequences of any places that are too dangerous for the children to visit. The visit might also prompt the setting up of a related mini-enterprise in the school.

MINI-ENTERPRISES

Although a visit to or simulation of an enterprise is very valuable, actually running a business in school involves children combining the elements of raising capital, designing and making a product, controlling stock, advertising and selling in real decision-making situations where the children can see the consequences of their actions. The children not only gain further insight into the structure and organisation of businesses, but they are also helped to develop interpersonal skills and are enabled to appreciate that different people can contribute different skills to succeed in a common goal.

A mini-enterprise can be set up by a small group of children or whole class and can be successfully executed by children throughout the primary school. Profit-making enterprises might include providing a service like car washing; making a product such as making and selling cakes and badges; growing plants; or being involved in distribution as in a tuck shop. The profits, if any, could be given to a charity, be used to buy school equipment, or even be distributed between the children. It is important to decide on what will be done with the profits before starting any enterprise, and perhaps more importantly how losses could be sustained. Non-profit-making enterprises can include organising an event such as a picnic, outing, play or concert, or a project to improve the local environment. The length of time the project will run also must be agreed and may be a one-off event, operate daily for one or two weeks, or take one day a week for half a term. The children need to ensure that health and safety for themselves and their product is satisfactory. For example they should ensure that the standard of hygiene during food production is adequate.

The actual project may arise naturally from an identified need in the school, follow from a visit to industry, or develop from inquiries carried out in the community. The children need to ensure that their ideas are feasible and that they are capable of carrying them out. For example one primary school set up an enterprise to produce and sell muesli with the support and advice of a local store. Initially they carried out a survey to find out what cereals were used at home. Recipes were designed and tested on the potential customers before the best were actually produced and sold.[13]

The children may need to raise money to buy raw materials. This might be quite a learning process in its own right as one infant class discovered. The children initially wanted to have a raffle but then realised they had no money to buy the prizes so did jobs at home in exchange for prizes. Other schools have borrowed money from the school fund, or raised a bank loan, and some even issued shares in their company and then paid out their profits to the share holders. Once the capital has been acquired, control of the money and stock is important, entailing some sort of bookkeeping and/or use of a spreadsheet.

Ideally every child should have an opportunity to try a range of responsibilities, and any stereotyping of tasks should be resisted. All the children should be able to have the experience of making, advertising, bookkeeping and selling their product, as well as taking part in the management decisions. On the one hand it is important that the children see themselves as genuinely making decisions; on the other hand the teachers need to be aware of progress so that they can provide advice and if necessary veto a particular action. Other adults, such as a bank manager, business director or trade unionist, might also visit to tell the children about their experience in business and give advice on handling money or safety conditions. The children are likely to find such visits particularly valuable because they can relate what is said to their own personal and practical experience of running a business.

PROVIDING APPROPRIATE PRACTICAL EXPERIENCES

Practical experiences (visits, role play, simulations and setting up mini-enterprises) need to be relevant to the age and maturity of the children. Early-years children can be introduced to ideas of buying and selling, processing raw materials, providing services and distribution through examining familiar concerns in the local community such as local shops, hairdressers, opticians, bakeries, farms and milk deliveries. In Key Stage 2, children can investigate less familiar and more complex establishments in order to understand mass production and the advantages of bulk buying.

Investigating local shops

A study of small local shops is an effective way of helping young children to understand that goods are bought, sold and advertised. It is probably easier to concentrate on establishments that sell a limited range of items and which will not cause conflict with children from different religions and cultures. The children could walk around the area around the school to see the different businesses and shops, one or two of which it may be possible to visit. One school arranged for several classes, including the nursery and reception classes, to visit different enterprises which were followed up by designing and setting up a role-play optician's, pet shop, bakery, supermarket, café and post office.[14]

A greengrocer's shop could be visited so that the children can find out where the fruit and vegetables come from and how the prices are set. Many children will think the stock comes straight from a farmer and imagine that the price the greengrocer sells the fruit is at the same price as it is bought for, or at a lower price because it is second hand. Such a visit could be followed up by setting up a role-play greengrocer's shop in which the fruits could be made of plasticine or dough. On the other hand real fruit can be sold at play-time, which can entail the children buying the fruit they are going to sell, working out suitable prices, labelling and displaying it appropriately to attract potential customers.

A shoe shop visit could be followed by the children setting up a role-play shop and stocking it with both old pairs of shoes and some they have made themselves. A mule-type shoe or flip-flop are fairly easy styles for the children to try to make by first making templates by drawing round the children's own feet. The shapes are then cut out from cloth and card and stapled or glued together. The children could try to price their shoes 'realistically' by looking at catalogues and noting the prices of actual shoes in the shop.[15]

A fish and chip shop or a fast-food take-away concern could be created after a visit. Again the actual design and establishment of the area lends itself to many technological activities, including designing a suitable environment, equipment and finding ways to produce imitation food from junk material, cardboard, tissue paper and dough or plasticine. Children can discuss how the establishment copes with staffing at unsociable hours, when there are extremely busy times, and at others when it is very quiet. This project gives opportunities to discuss staffing costs and the need to control stock carefully as, if too much food is bought, unsold perishable food items may have to be thrown away and, if too little is purchased, food will run out and the customers will be dissatisfied.

A hairdresser will provide an example of an industry where payments are made for a service and no tangible item is taken away. The children can try to discover why a charge is made at all. They should find that hair-products, wages, advertising and rates have to be paid for. If they compare prices of different establishments, perhaps by asking relatives, they are likely to discover that these vary quite considerably, often because of the spending power of the customers. Some establishments, for example, concentrate on catering for old age pensioners who are very regular customers and who ensure that the shop is nearly always full but who are unable to pay high prices. Other places which cater for a more fashion-conscious but irregular clientele may charge more. The children could collect examples of different hairstyles for men and women from magazines to display in their 'shop' and could discuss how fashion changes cause the hairdresser to have to make adjustments to the equipment, hair products and displays.

A garden centre visit can lead to many simple scientific experiments on optimal growth conditions and might also lead to the children setting up a mini garden centre in the school which can encompass raising capital for the seed, pots and compost and later selling the plants. Such a project should help the children to realise that the production process also entails tending the plants, giving them time to grow, finding a market, advertising and setting realistic prices. The children may also have to come to terms with the problem that elements outside their control, such as wind, rain, frosts and disease, may affect the success or failure of the project.

Introducing the idea of processing raw materials

As part of the Geography National Curriculum, children at Key Stage 1 are expected to consider where common materials are obtained.[16] This study can be effectively linked with technology when the children are introduced to the idea that these raw materials need to be procured, processed, distributed and sold.

As with a garden centre, the children should be able to have some sort of practical experience relating to most of the stages of producing bread from raw material to the outlet. Some wheat, barley or rye seeds could be obtained and grown in pots in the classroom. The children could try grinding their own flour from cereal wheat grain from a health-food shop using a rolling pin, which they will discover is very hard work, leading to an appreciation of why most flour is ground by machine between huge rollers. Examples of several types of flour could be examined and different kinds of bread (pitta, roti, pizza, wholemeal loaves or buns and white bread) made in the classroom. Once the bread has been made the children can complete the simulation process by slicing and wrapping it. There may be a bakery, perhaps in a supermarket or department store, which the children can visit so that they can compare the differences of mass production with their individual items made in the classroom.

In order to examine the stages of manufacture of a wool pullover or scarf it may be possible to acquire some sheep's wool, ideally as part of a visit to a farm; alternatively animal wool can be bought from some chemists. This could be cleaned and dyed and the children could try to spin it.[17] They might also try weaving and knitting. Of course not all clothes that appear to be wool are made from natural materials, and by looking at labels in their clothes the children could sort out those made from wool, cotton, etc. and those made from man-made fibres such as nylon and acrylic, which could lead to talking about other ways of producing clothes. As the final part of such a project a role-play clothes shop might be set up in the classroom.

Work started in the early years can be extended by studying different products and in greater depth. The children could look at how they get a particular item such as milk, a toy, meal, or television and trace back its origin. Many such studies of where raw materials come from and how they are processed will rely on using books and other second-hand material, but if at all possible the work should be linked with visiting an appropriate business or industry.

Transport and distribution

Transport and distribution are important elements in most production processes. Early-years children might watch the local traffic to see what different delivery vans are carrying. Children with families who have had the experience of living in different parts of the world may be able to add how goods are delivered to homes and shops in those localities. The children could try to find out where the lorries and vans that call at the school have been, what they are carrying and where they are going in order to help them to see that all the food and resources in the school have been bought and are being delivered from a factory, shop or depot of some type.

Older children might collect packages and labels from different items to find out where they come from. The labels from tins, such as fruit and vegetables, could be displayed around a map of the world linked to Britain with threads to show which countries they have come from. From the labels the children may also be able to find out whether the raw materials were processed in the country of origin before being transported or afterwards. From this information they could mark the position of British food-processing industries and other factories on a map of Britain with lines to show that the goods have been transported to the children's home area. The children could also investigate how different materials are carried and how the transporters, such as container lorries, tankers, cargo ships and goods trains, are designed for their particular purpose.

EXAMINING PRODUCTION FACTORS

Older children can be helped to understand that the cost of production includes time, people, skills and materials. They should learn that goods can be made singly or in quantity as on a production line but this affects what the item costs. A number of exercises and simulations can introduce these ideas to children so that they have a better basic understanding when they visit unfamiliar and/or complex industrial concerns.

Production costs

A way of helping children appreciate that the use of money is related to buying materials, skills and people's time is to ask the children to make a fairly simple artifact which can only be made by buying resources and assistance with a limited number of tokens. These could be buttons, conkers, cocktail sticks or imitation money. The children might want to make a moving puppet made out of cardboard and split pins or a tip-up lorry from junk materials, balloons, tubing, glue, sticky tape and wooden wheels. All basic materials are pre-priced and have to be paid for, as does any help required from teachers and parents. The teacher will have to decide what to do with those children who use all their 'money' before completing the item. This experience should help the children to appreciate the importance of the cost of materials, the restrictions of limited resources or capital, and to a certain extent the need to budget to buy people and skills.

Following a study about the production of a commodity or any visit to a retail establishment, older juniors could try a drama or simulation of how money is used to buy the raw materials and pay for wages and delivery. Two producers, transporters, wholesalers, shopkeepers and customers are needed. Each player has some money with the exception of the producers who have all the goods (e.g. represented by egg boxes). The customers are given regular pay packets. The idea is to get the produce to the customers but with each person in the chain getting some profit so they can eat and look after themselves and their families. The price the goods change hands at is determined by bargaining. By having two of each category of occupation an element of competition is introduced. Customers, as well as others in the chain, may bargain, as in some countries this is quite normal and equates with the fact that in Britain the customer may choose to use a different shop if one is found to be too expensive. Initially the children find the activity quite difficult; as they are so keen to sell their goods they may do so at a loss and then demand more money when they go broke.

Bulk buying

If materials can be bought in large quantities they are usually cheaper. The children could compare prices of the same item, for example cereals or washing powder, being sold in different quantities to find out what size would be the best buy. Another exercise to compare bulk production with individually made items could involve all the children making different badges to sell at play-time to compare the ease and cost of everyone making one class design. When the children make the individual designs, perhaps advertising the joys of reading, or on an environmental issue, they will probably find they have used a wide range of materials and tools; wastage has been high; and the assignment has taken quite some time to complete as it involves designing, checking availability of materials and correcting errors. The children are also likely to find that some badges sell well and others do not. One design could then be chosen for mass production and made as on a production line. After the exercise the children could discuss the differences and relative merits of each method. They may comment on the advantages of enhanced skills in performing a single process compared with boredom and job satisfaction. They might also be able to see that by having only one design, time for the production process is quicker; fewer tools are needed; stock control is easier as it is easy to see when a component is getting limited; and large quantities of a limited number of categories of raw materials means that they can be acquired by bulk buying. The children may be able to relate this experience to explain the differences of cost in designer clothes compared to those sold in a chain store or handmade furniture compared to mass produced items.

Production lines

An extension of the economies of bulk production is to streamline the assembly process by a production line. Making construction-kit models, paper cars, party decorations, birthday cards and imitation flowers can be set up as production-line simulations. A conveyor belt can be made by placing a number of tables in a line with a strip of old wallpaper twice the length of the tables placed along the tables with the two ends joined underneath. Several children can be given a particular sub-assembly task in order to complete a fairly simple constructional model. The children are sat facing the 'conveyor belt' and as it is moved each child adds their parts to assemble the model as it goes past. They could try the consequences of moving the conveyor belt slowly or quickly.

A similar arrangement can be made to simulate a car production line to assemble a side-view two-dimensional paper car made of pre-cut-out shapes for the body, roof, two doors and two wheels. The first person glues the body to a piece of card. This is placed on the paper and moved to the next person, who sticks the roof on. As the car moves along it is built up. The simulation might be extended by putting the children into groups to form separate companies with specific roles such as manager, workers and quality controllers to produce model cars or imitation flowers made of tissue paper and pipe cleaners or wire. In one school which tried a similar simulation the 'managing directors' even introduced the idea of supervisors and bonus schemes.[18]

IDENTIFYING AND INFLUENCING CUSTOMER NEED

Market research

There is no point in producing a commodity or giving a service unless there is a need or market for it. Early-years children need to appreciate that everyone does not have the same needs or preferences and they should use this knowledge to design and evaluate products. Older children should have opportunities to appreciate that these preferences may result from different values, cultures and beliefs. Additionally they should also learn that the product quality, cost and appearance influence consumers.

Very young children might first compare their tastes with those of their peers by talking, drawing or writing about their favourite toy, game, television programme, sweets or clothes. They could also be asked to say how they would spend 50p and explain their choice in order to compare different children's preferences. In order to link the idea of asking questions about personal preference to a subsequent design activity the children could be put into pairs and be asked to draw their partner's ideal party clothes. This will require them asking what their friend likes rather than concentrating on what they themselves want. As part of setting up a role-play grocer's shop the children could make and test different types of sweets, biscuits and cakes to record how preferences vary within their own class, or they could make some to sell at play-time and record which their customers buy first and particularly like so that these could be made on another occasion.

To collect accurate information about other people's views, more formal methods of gathering and processing data are needed. In order to improve the buying policy for a tuck shop the children could design a questionnaire to discover which crisps or snacks are most popular throughout the school so that these could be ordered in greater amounts. The data can be analysed and presented in different graphical forms by hand or by using a computer database. The questionnaire can be analysed and improved by asking questions such as: Were all the questions unambiguous? Are open-ended or multiple-choice questions better? Were the categories of the multiple-choice questions discrete and comprehensive? Did the respondents understand all the questions? Was the layout understandable? Were the data easy to extract?

As adults usually have very different preferences to children, this can be explored by drawing up a questionnaire to discover what television programmes they and their families watch in a week, which are particularly popular with children and which are mainly watched by adults and why. The children could then suggest how the television programming could be improved. The teacher may wish to discuss the constraints of the television companies which have limited budgets and have to try and cater for many tastes.

By the end of Key Stage 2 children should be increasingly able to consider the needs of people other than their peers or well-known people in the community and realise that these individuals' views will be influenced by their age and culture. The children could visit an old people's home to talk about the residents' lifestyle and what interests them. One group of children discovered that the old people liked to play cards but some with arthritic joints had problems holding them. This led to a card holder being designed and made for them. Visitors could be invited to the school to discuss different cultural needs or the requirements of the disabled to help children firstly appreciate the lifestyles of different people and then to try to find out what might make their lives better.

People are not only influenced by their basic needs, cultural requirements and cost; other factors such as convenience, appearance, ambience and of course fashion are important. The importance of different factors could be investigated by comparing families' use of different shops. In order to find out whether the prices vary much between the shops the children could choose about 10–20 basic food items and then find out the cost of these when they subsequently visit the shop with a view to discovering which establishment has the best range and value. The quality of the goods can also be discussed as the cheapest brands might not necessarily taste best, last as long or be as reliable as more expensive items. A survey can then be designed to ask parents what shops they use and why in order to discover whether price was the significant element or whether other factors were more important, such as convenience, range of goods, quality of the food, or friendliness.

In order to keep consumers better informed about different products some magazines collect information about different products by testing them and surveying existing owners or consumers so that they can inform potential customers about the best buys. If the children look at these magazines they will see that the recommended best buy is not necessarily the cheapest as quality, durability and aesthetics are also significant. The children could write their own reports which could include evaluating different pens for the quality of line they produce, i.e. even and continuous; how they perform when colouring large areas; how comfortable they are to hold; whether they dry up quickly if left on a radiator; and what happens if the tops are left off, as well as cost and aesthetic design. The children could also look at different products such as natural dyes (e.g. red cabbage, beetroot, lichen and coffee) for the colour quality,[19] or floor surfaces for a kitchen (e.g. tile, cork mat, vinyl square, rubber mat and carpet tile) for their ability to be cleaned and withstand damage if scratched or when things are dropped on them.

Fashion is another significant influencing factor on potential customers. The children can discuss occasions when they are or have been influenced by fashion, e.g. toys linked to current television programmes, recreational fads such as skateboarding, sports clothes with particular logos, and tapes and CDs of pop groups. These industries, in particular, need to carry out market research on their customers' preferences and likely financial limitations, as well as be very active in advertising to both inform their possible customers of their products and also to influence fashion and create need in the minds of the potential buyers.[20]

Advertising

If a business can identify the market for a product accurately and then direct their advertising at this group they are more likely to succeed. Advertising enhances choice by providing information but it may also distort the truth by, at the very least, concentrating on only the good features of the product or service.

As an introduction to advertising the children could be given a page advertisement from a magazine to compare with a short factual article in order to try to identify some of the special features of advertisements. They might try producing two equivalent pages for a magazine to advertise a very familiar feature, such as the school or local park, and to write a short factual piece about it. The discussion and evaluation generated by the activity should help the children to start to appreciate how advertisements concentrate on only the positive aspects and ignore disadvantages of the product; the significance of the initial visual impact; and the importance of the language used. They may be able to start discussing ethical considerations and the value of having controlling bodies such as the Advertising Standards Authority and how people may be persuaded to buy products or may pressure their parents to buy things that cannot really be afforded or might not be wanted in the long term.

The children could list the places where they find advertisements (in magazines and comics, and in posters on the street or buses, on television and radio, through the letter box, and in the classified columns in newspapers) and a collection made of different types of adverts to investigate. It is relatively easy to collect a variety of adverts from magazines for evaluation. Initially the children will be able to comment more effectively on those directed at them, which will be found in comics and magazines for young people. They might choose their favourite and ask questions such as: Who is the advertisement for? What does it do to attract the reader's attention? What are the main points? How much print is used and how? What colours and pictures are used? What design features are good and which are poor? How could it be improved? Is it their favourite because of the advertisement's design or because they like the product, or both? The children might also take two different advertisements for similar products, compare them and try to identify what makes a successful advertisement. Such exercises offer excellent opportunities to develop evaluative skills as well as help the children to improve their planning and design.

By looking at a wide range of different magazines directed at fairly specific audiences, such as ones for people particularly interested in fishing, horse riding, motor cars, cooking, clothes and pop music, the children can compare the different advertisements and will discover that most are targeted to the group reading that magazine. The advertisers must not only know which people are likely to be interested in the product, they must also make the design appeal to that group and then place the advert where these people are most likely to see it. Subsequently the children could produce their own advertisements directed at different audiences, perhaps one advertising the sale of soft drinks for the early-years children, and one to advertise a class assembly for parents. The project may involve market research, such as finding out why the young children would want a drink at play-time, what drinks they particularly like, what words they find easy to read, and where they are most likely to see the posters.

Questionnaires to find out about how television advertisements appeal to different groups and which are broadcast at different times can develop these ideas further. The children could choose to write about their favourite television advertisement and try to suggest why it has a particular appeal for them. Is it because it is clever, makes them laugh, has a story line, is about people they would like to emulate, has appealing music or because the product is particularly good or fashionable? As a comparison a questionnaire could be drawn up to find out what appeals to different people in their families. The children could ask relatives what their favourite and most disliked advertisement is, and why. A survey could also be carried out to discover whether television advertisements are targeted at different groups of people by listing what is advertised during children's programmes compared to those directed at different audiences such as the news, current affairs programmes, soaps and documentaries.

By looking at one or two recorded advertisements in the classroom the children might also look at the techniques used by the producers to help the audience remember the product, such as the numbers of times the name of the product is mentioned, whether there is a particular catch phrase, and how often the logo or actual product is seen. It might also be possible to discuss the use of stereotypical or idealised situations so that the audience is encouraged to identify with the actors but with the implied promise that with this product life is better. For example families are rarely shown with very messy kitchens, screaming babies or quarrelling people. The children might also consider how people with disabilities and those from different races and cultures are depicted, if at all.

Following these experiences groups of children could write a storyboard and finally produce short videos advertising different imaginary products. Each group could be given a different challenge: perhaps an educational toy for nursery children; a range of non-fiction books directed at seven years olds; a lunchtime computer club; or a range of T-shirts.

The study of classified advertisements should help children to realise that advertising costs money. The children can be challenged to write adverts for different items using the minimum words, but giving the essential details, within a limited budget, and then work out the value to the paper of a column of advertisements. This will help them to realise how some free papers fund their production. In the same way much of independent television is funded by revenue from advertisements. Consequently industrial producers have to make careful decisions about how much they can afford to pay for advertising campaigns and need to carry out very careful market research to ensure their money is not wasted.

Methods of informing and influencing the public are not confined to advertisements, as most products that are sold are packaged in a way to attract the potential buyer as well as protecting the contents. By collecting different boxes and containers the children can evaluate manufacturers' packets as well as attempt to identify what factors need to be considered when designing packaging in order to make containers of their own. Most packets basically need to protect the contents in some way: to preserve them; stop them from being crushed, wasted or spilt; and to keep them clean; and the shape must make effective use of display space. However the packet must catch the customer's eye and then persuade him or her to buy by having an interesting shape or exciting design. Until recently many packets were bigger than their contents required so that the customer was given an impression of good value for money. However recent increased environmental awareness on the part of the customer has encouraged some manufacturers to promote their products as being ecologically friendly by being small, concentrated and recyclable.

Food packets, by law, need to have details about the ingredients, food values and chemicals, as well as dates indicating by when the food should be used. Packages may also have instructions for use, construction, safety, and some form of bar coding to assist stock control and the checkout process. All this information must be provided in such a way as to be attractive and entice the customer, or be included in such a way as it will not interfere with the overall image. The children can study the way packet design employs pictures of the product and its uses, logos of the company, colours and print size to attract the buyer as well as providing all the necessary information. By taking boxes apart the children can work out what nets are required to make similar boxes so that they can design and make boxes to sell cakes, biscuits, sweets or other items that they have produced.

The use of logos is a way of constantly reminding people of a business. A logo needs to be simple so that it can be reproduced easily in many different locations, such as on letter headings, packaging, clothes and notices as well as major advertisements. The effective logo should be distinctive and convey the name and image of the product. A collection of well-known logos could be made from major chain stores, food outlets, sports clothes, sweets and public services. The children could try these out on different people to find out which are easily recognised and which liked or disliked. A class logo might then be produced. The children need to decide what image they want: studious, sporting or friendly, perhaps; and what icon might portray this: a book, running figure or smiling face. They need to decide on the colour and shape and may be able to produce it on computer so that it can be easily reproduced.

A whole technology project could be centred around an advertising campaign. One junior class planned and carried out an advertising campaign for an imaginary clothes shop. The children decided to design and make posters, magazine advertisements, hats and T-shirts and planned a special programme for the opening day. The project successfully covered many technological skills as well as developing the children's English and art skills.

Issues of safety and environmental impact

It is important that the organisation of the enterprise ensures adequate safety for the workers when they handle equipment and potentially dangerous raw materials; sufficient rests and breaks must be provided; and the working environment should be appropriately heated and lit. The safety of the customers must also be considered. For example food must be hygienically made; toys should be safe for young children to put in their mouths; high chairs need to be stable enough for wriggling babies. When children make visits to different enterprises they might be asked to look for potential danger areas; ask what rules there are for safety and who makes sure that they are followed; and discover what safeguards and tests are carried out to ensure that the product is safe for the consumer. Subsequently the children could produce posters to remind people of safety rules both in the business they have visited and in their own school.

Industry must also consider the effect of their enterprise on the environment in general. Does the product use unnecessarily large amounts of limited resources such as wood from the tropical rain forests and unrenewable energy supplies? Could these be used more economically or alternatives used? Will the product itself cause ecological problems such as in sprays that contain CFCs and cars that use lead petrol? Could a better product design avoid these disadvantages and would an appropriate advertising campaign persuade the consumer to pay more to cover any increased production costs? Is waste produced by the industry carefully controlled? As industry is becoming increasingly aware of its responsibilities to protect the environment the children could explore these issues with some of the industries they visit or they might try to find out about products which claim to be environmentally friendly in order to assess the justification of their claims and compare their costs and quality with alternatives.

By enabling children to become more aware of factors involved in providing a service or product as a business enterprise they will start to appreciate industry's potential as well as the constraints and limitations. Hopefully their experiences will make them more discriminating customers and effective and responsible entrepreneurs in the future.

NOTES AND REFERENCES

1   Durowse, H. and Elliot, M. (1987) 'Foreword' in Smith, D. (ed.) (1988) *Industry in the Primary School Curriculum: Principles and Practice* Lewes: Falmer Press.
2   Lewin, R. (1981) 'Technology Alert! – in the primary school'. *School Technology* Issue 60 Vol. 15 No. 2, pp. 2–4.
3   Fitzpatrick, S. (1988) 'The "We Make" Projects, Forsbrook Infants School, Stoke-on-Trent' in Smith op. cit.
4   Jahoda, G. (1979) 'The construction of economic reality by some Glaswegian children' *European Journal of Social Psychology* Vol. 9, pp. 115–27.
5   Linton, T. (1990) 'A child's-eye view of economics' in Ross A. (ed.) *Economic and Industrial Awareness in the Primary School* London: PNL Press.
6   Schug, M. (1990) 'Research on children's understanding of economics: Implications for teaching' in Ross op. cit.
7   Linton, op. cit.
8   Berti, A. and Bombi, A. (1981) (English translation by Duveen, G. 1988) *The Child's Construction of Economics* Cambridge: Cambridge University Press and Editions de la Maison des Sciences de l'Homme.
9   Ross, A. (1990) 'Children's perceptions of hierarchies in industry: A developmental model' in Ross, op. cit.
10 Smith, op. cit.
11 Ross, A. and Smith, D. (1985) *Schools and Industry (5–13): Looking at the World of Work: Questions Teachers Ask* London: School Curriculum Industry Project gives useful general advice for teachers exploring possibilities of using industry to develop their curriculum.
12 Jamieson, I. (1985) 'Industry and the primary schools' in Jamieson, I. (ed.) *Industry in Education* Harlow: Longman.
13 Baker, T. (1991) 'Co-op co-operates with education' *Design and Technology Times* Autumn, pp. 8.
14 Hotspur Primary School Heaton, reported in Ross, A. (1989) *The Primary Enterprise Pack: Eight Schools* London: PNL Press.
15 Petty, K. (1991) *New Shoes* London: A & C Black illustrates the stages of production involved in developing a new range of shoes.
16 DES (1991) *Geography in the National Curriculum (England)* London: DES/HMSO.

17　Dixon, A. (1988) *Wool* London: A & C Black is a children's book with clear pictures of the production process and activities for children to try.
18　Fitzpatrick, op. cit.
19　Richards, R. (1990) *An Early Start to Technology* Hemel Hempstead: Simon & Schuster.
20　Petty, K. (1991) *New Shampoo* London: A & C Black illustrates the production of a new shampoo with useful sections on market research, packaging and advertising.

FURTHER READING

Benfield, E. (1987) *Industry and Primary Schools* Cambridge: Hobsons.

Fulford, J., Hutchings, M. and Ross, A. (1989) *Bright Ideas: World of Work* Leamington Spa: Scholastic.

Hutchings, M. (1989) *The Primary Enterprise Pack: Children's Ideas about the World of Work* London: PNL Press.

Jamieson, I. (ed.) (1984) *'We Make Kettles': Studying Industry in the Primary School* York: Longman.

National Curriculum Council (1990) *Curriculum Guidance 4: Education for Economic and Industrial Understanding* York: NCC.

Ross, A. and Hutchings, M. (1990) *Enterprise, Economic and Industrial Understanding Kit* London: PNL Press.

Waddington, D. (ed.) (1987) *Education, Industry and Technology* Oxford: Pergamon Press.

Chapter 9

Developing effective cooperative groups

Discussions and collaborative work are an important aspect of design and technology.[1] Children should work with others in planning and apportioning tasks, being prepared to explain, clarify and justify their ideas to others. They also need the experience of working in groups so that they can develop leadership skills and learn to understand, design and develop organisational systems where each person has a specific role. In addition, interaction with others in a group who hold different levels of understanding or skills can help children to develop a greater understanding of many of the concepts and skills required in technology.

While grouping is a widely used organisational strategy in primary schools, the deliberate and purposeful use of collaborative group work, where a number of individuals work together to achieve one outcome, is much less common. Traditionally, throughout the curriculum, children have rarely been given opportunities to work on group tasks partly because, before the National Curriculum, there was no specific demand on children to work together. A large-scale observational study of primary-school classrooms before the implementation of the National Curriculum found that 69 per cent of teachers never used cooperative group work for art and craft or for topic work, perhaps the subjects mostly closely related to design and technology, and even less in other subjects.[2] By sitting together, children do not automatically debate, assist each other or share ideas, as the study uncovered. In the groups that were observed, individuals spent two-thirds of the time working on their own, talking with no one. Only 5 per cent of the total time was spent talking to other pupils in relation to work and most of this was directed to children of the same sex. A later study with infant children reported more task-related talk between children but it was of a low order.[3] There were few instances of explanations, discussions, decision making and demonstrations which furthered the successful completion of the work. It was also noticeable that interaction was marred by the fact that the children were often confused about whether they should be cooperating and what type of mutual assistance was permitted. Therefore, to develop true cooperative groups, children need opportunities to work on shared projects where they are clear that collaboration is required.

ADVANTAGES OF COOPERATIVE GROUPS

Dunne and Bennett[4] investigated cooperative grouping with teachers who had not attempted to develop this approach before. The teachers tried language, mathematics and technology activities. Without exception they found it easier to implement than they had imagined. They commented on a noticeable increase in discussion, suggestion, testing, inferring and giving informed conclusions by the children and reported that the work was more thorough and better presented. The teachers also found that it gave them more time which could be used for assessment or to teach more intensively. Cowie and Rudduck,[5] working with secondary teachers, reported that committed users of cooperative group work considered that the children's skills and competence in communicating increased as they became better able to express a point of view. They felt that the children learned to share ideas, built on others' contributions and developed a sense of audience. The pupils became more sensitive to the rights of others to contribute, valued what they had to say, and showed that they could identify and use the resources of the group to advance their own learning. The pupils were also thought to develop their ability to apply different concepts, test hypotheses, interpret and analyse data, and relate their learning to their own experiences. They were reported to have increased self-respect and became more responsible for their own and the group's work. Research in the USA supports these findings, indicating that cooperative learning promotes higher achievement and self-esteem, and promotes increased positive attitudes to school and classmates irrespective of race or ability.[6]

Although there is real value in promoting cooperation within groups there may be a case for having competition between groups as a technique for maximising individual effort within each group; the limited work investigating this question, however, indicates that the children perform better without such competition.[7] Some competitive activities will no doubt stimulate the class, perhaps as end-of-term projects, but it seems advisable to concentrate on developing cooperation in the class as a whole, as well as within groups, in order to be able to discuss and establish the necessary social and problem-solving skills for effective teamwork with all the children.

Of course not all technological activity should be in groups, as there is a place for whole-class and individual teaching, but it is a significant method of delivering this subject. When organising this group work it can never be assumed that it is enough to divide the class up, announce the activity and leave individuals within the group to interact purposefully. Effective group work depends on pupils developing appropriate skills and attitudes, and on careful preparation and meticulous management.[8] Therefore careful thought needs to be given to the size and composition of groups, methods of organisation, and strategies to help children to develop cooperative skills.

GROUP SIZE

The fewer the children in the group the more limited is their opportunity to learn how to work with a variety of people. On the other hand individual children in large groups may not have enough chances to speak or provide a significant contribution to the product. Initially young children and those unused to working in groups are probably better put into pairs with the aim of building up their expertise of cooperative work. Subsequently group size should be varied and primarily determined by the nature of the assignment and the number of natural tasks with it, as too many children allocated to an activity which is limited by the working space, equipment and sub-products will result in some children being under-occupied. Even given an assignment which has many activities that could be shared, experience indicates that groups of five or six break down in practice into pairs and threes, indicating that groups of four are usually the optimum size.

GROUP COMPOSITION

In design and technology there is a case for using a wide variety of grouping strategies according to the type of activity and the social and intellectual needs of the children, including friendship and single-sex groups; children of similar ability or attainment; mixed ability; and groups chosen with regard to the children's personalities.

Friendship groups

When an activity is very new, or could be emotionally stressful, it is often advantageous to use friendship groups so that the children feel at ease with each other and give one another moral support. If children who have had little experience of dance are asked to design and prepare a dance for the first time, they are more likely to succeed if they can choose whom they work with. In a similar way, open-ended problem solving and role-play drama are initially best introduced to the children in small self-chosen groups. As the children become more secure in the approach of the task the teacher might wish to take other factors into account.

Gender

Lack of confidence may also be a valid reason for occasionally organising the children in single-sex groups. Valerie Morgan[9] tried different groupings in science activities, where research indicates girls are less assured than boys,[10] and found that groups which consisted of one girl and three or four boys almost always led to the girl withdrawing. Even when there was one boy and three or four girls the boy was often able to exert strong influence and get a disproportionate access to the equipment. In the all-girl groups, however, the girls seemed most secure, and less afraid of trying things and putting their ideas forward. There is a case, therefore, in sometimes having single-sex groups in technological activities particularly closely related to science where there is a risk that the boys will dominate discussions, activities and equipment. A similar argument might be made for those technological projects where boys may traditionally lack experience until their confidence and competence have been established.

Most friendship groups are likely to be single-sex groups, particularly in older classes and schools with a high percentage of children from Asian and Islamic cultures. Therefore an over-reliance of friendship and single-sex groups is not satisfactory, as children will meet mixed collaborative groups as adults and should have the opportunity to develop skills of working effectively with people of a different sex. In addition, all-girl groups tend to want to seek agreement and reduce tension rather than probing and challenging each other so perhaps they do not always develop their ideas as much as in mixed groups.[11] In technological activities where skills and concepts are fairly well established, groups should be frequently mixed. This process will be assisted by an active whole-school policy promoting the view that all technology, as well as all other subjects, is of equal interest to boys and girls.

Grouping by ability or attainment

It must be remembered that levels of ability in mathematics, English and technology are unlikely to coincide exactly. It has frequently been found that children considered to be in need of remedial help in reading or number work show real ability in investigative and practical work, and some very academically able children find the creative, open-ended and imaginative thinking needed in technology difficult. Therefore ability grouping in technology must relate to the concepts and skills of this subject. These ideas can be ascertained by drawing, concept mapping or through practical work and discussion. Teachers will also be able to use the increasing amount of information becoming available on the way children's concepts develop in areas like forces, energy and economics in order to assist them to categorise and group the children.

Bennett and Cass found that high attainers do well in both ability and mixed-ability groups and do not appear to be disadvantaged by being with low attainers.[12] However their research indicates that ability grouping for cooperative activities is not as effective for low attainers as for high attainers, as the former generally do not have the level of knowledge and understanding to provide each other with support. The teachers also found that they needed to spend more time with these groups in comparison with cooperative groups composed of mixed abilities and able children. As a general rule, therefore, grouping by levels of attainment is probably not appropriate for most cooperative technological activities. However teachers may decide on occasion to have an ability group in order to provide specially tailored activities to meet particular needs or in order to spend time with the less able children to help them to develop cooperative skills.

Mixed-ability grouping

High attainers are of particular value in mixed ability groups where they are able to support the less able children with inputs of knowledge as well as with suggestions and explanations. However the proportion of high attainers in the group is likely to be significant. It was found that when two high attainers were grouped with one low attainer the latter tended to be ignored or opted out. When the proportions were reversed all the children performed well.[13]

When the collaborative task involves sharing ideas about science or economic concepts (e.g. relating to transfer of energy or forces) it appears there are positive advantages in having mixed groups. As the children discuss their ideas they are able to rethink their incomplete or incorrect ones to develop ideas more acceptable to current scientific and economic thought and so make greater progress than groups composed of children with similar concepts. This improvement appears to be as significant in both children holding more advanced concepts as well as those with a lower attainment.[14]

In mixed groups the teacher must be aware when children expect leadership from one particular child; others may be hesitant in putting their ideas forward, and good ideas from younger children and those with low status may be ignored. If this occurs the teacher may need to point out that everyone's ideas must be shared and as technology involves such a wide variety of skills all of them should be able to excel in some aspect. Indeed children with special expertise, such as those who have a good grasp of using a particular word-processing program, database or spreadsheet, understand how to convert a recipe to cater for different numbers, or are able to use woodwork tools correctly, can be placed in mixed-ability groups to assist the others.

Personality

American research has identified three types of behaviour that would interfere with effective group participation, i.e. 'free riders', 'suckers' and 'gangers'. 'Free riders' tend to occur if there is an imbalance in the rate of work by the individuals in the group. Children may opt out or go through the motions of cooperation because they feel the group is being held back by having to wait for someone who is slow; because they want to spend more time on a high-quality finish whereas the others wish to complete quickly; or because they find it difficult to cope with the application of advanced concepts. The teacher can avoid many of these instances by varying the group composition so that projects which consist of very similar tasks are given to similar ability groups, and mixed-ability grouping is used for projects with a wide range of sub-tasks. It is also important for the teacher to acknowledge these feelings of frustration and by doing so the 'free rider' often feels able to rejoin the group. 'Suckers' are children who initially throw themselves into a task and work hard to develop the product but become demotivated if the others in the group appear to be using them and taking them for granted. This is less likely to happen in small groups and where the size of the group has been chosen to match the number of natural tasks. In the third situation when children see little point in the activity or who do not like the type of work, they may 'gang up' on the task to find ways round doing it. This problem can be partially avoided by involving the children in the decision-making process of deciding what products to tackle and how they might go about it.

While attempting to maximise each individual's opportunities within cooperative work it is more difficult to cater for different personalities than to adjust group membership to match the task. Pupils who are naturally dominant, appear to enjoy being argumentative, are so subservient in their desire for acceptance that they do not offer ideas of their own, and/or have a real talent for technology but are not able to change their peers' expectation of them as generally less able, can create difficulties for themselves and their group. Mismatch of personalities cannot be solved overnight, but by actively teaching about cooperative group skills and encouraging self-evaluation children can be helped to become aware of how their behaviour affects others and how they might adjust it.

It can be seen that there is no easy way of deciding group composition. There may be a case on one occasion to have only friendship groups, and on another very carefully chosen groups which might in part be mixed-ability groups with one or two ability and single-sex groups. However carefully the groups are chosen there are always some children left over who do not fit easily in the proposed groups, which is an additional reason for varying the group composition so that at some time each child has opportunities to work in a situation designed for his or her needs. In whatever way the groups are chosen the reasoning behind the choice could be shared with the children, with the opportunity for them to put their points of view and suggestions. In some cases this might be done before the activity, and in others as part of the evaluation. The children are then in a better position to appreciate that the teacher considers group work to be important as well as to appreciate some of the aims for collaboration.

TYPES OF COOPERATIVE GROUPS

The type of group organisation will depend on the activity and experience of the children. Young children and those with little experience of cooperative activities need to work in fairly simple arrangements organised by the teacher.

Increasingly the children should become more autonomous, making their own decisions about their organisation and products with minimal input from teachers.

Encouraging cooperation between individuals working on identical individual tasks

It is easiest to get cooperative behaviour when the children have to create a joint product which cannot be completed without input from all concerned. However there are activities that cannot be easily shared by even two children where each individual needs an item, such as a gift, to take home or as in the case of a miniature garden when the product is small with few elements. There are also occasions when the teacher will want every child to do a similar task, especially when the main aim of the session is to teach a skill or concept. For example, in order to teach one method of using a lever or cam, the teacher may wish every child to make a pop-up toy, which can be customised but has the same basic structure and method of construction. While they are working individually it is then possible for the children to share the experience and contribute to each others' interest, skills and motivation. However, unless the teacher actively promotes cooperative behaviour, it is unlikely to be very noticeable. This might be done by the teacher informing the class that he or she is particularly looking not only for correct use of a lever, but also for children who share ideas and advice. The children's awareness of cooperation is thus heightened by being told that it is valued and will be assessed. An additional shared element could also be added to the main task by asking each group to make an agreed list of difficulties and points to remember when tackling a similar task in the future.

Individuals working on discrete tasks leading to a joint outcome   Year 6

As young children initially can find close cooperation difficult it is useful to give them a group task where there are obvious discrete elements which can be done individually. The overall task might be to produce a puppet show or musical performance in which individuals need to make a puppet or musical instrument. Once the group has discussed the task, shared ideas and agreed responsibilities, perhaps with the support of the teacher, the children can then choose to work alone. In this approach the children feel secure, being clear about their responsibility and not being threatened by having to compromise unless they wish to. However they are motivated to cooperate and support each other when problems arise, as the group product cannot be achieved until everyone has successfully completed their piece of work. This is also a useful approach to introduce children to the idea of sharing out tasks and coordinating their activities.

Joint work on a common product

As the children become more experienced in working cooperatively or when the task is relatively simple, once the problem has been established the children can be left to solve it themselves with minimal teacher assistance. Groups could be asked to find a way of keeping tidy, but easily accessible, the pencils, rubbers and pens that litter the teacher's desk; design and make maps and notices that can guide a visitor around the school; read about a country in order to produce a travel agent's brochure; or create an advertisement on video. If there is an open-ended element in the project, alternative strategies will need to be considered and once an approach has been agreed the children will have to share out and coordinate tasks and perhaps appoint a group leader. The same problem might be given to several small groups so that there can be a final review and evaluation of different solutions. As it can be difficult to ascertain what each group member has contributed, part of the evaluation process might include individuals reviewing their particular input.

Sub-groups working on tasks to achieve a joint outcome

A further step is to help small groups of children negotiate and share tasks that will contribute to a very large group or even a whole-class product. An early-years class may decide to design and make a fairground for small dolls. The teacher will need to guide the discussion, help the children come to a consensus about the end product, and help them sub-divide the task such as making a big dipper, helter-skelter, roundabout, ghost house and café. Each small sub-group can then be left to organise and solve their particular task for themselves but to regularly share progress and problems with the teacher and whole class. Where the organisation is complicated and involves the whole class, even the oldest children who are experienced in working cooperatively may need the teacher to take an active role in the organisation, major running decisions and whole-group review of progress. For example this might be appropriate if the class arranges a sports day with different groups taking on responsibility for designing, preparing and organising various events; or produces a magazine with the groups researching topics such as current affairs, sports, advertisements, cartoons, cookery and fashion, supported by editors, artists, layout artists and a production and distribution team.

Specialist groups to research projects

An important element in producing a design proposal is to research the project and to make tests of possible materials and ideas. One way of encouraging the children to do this is to instigate specialist groups in addition to the normal production group. For example, the children might have been divided into groups of three to design clothes for visiting a farm in winter. Each one of the three also belongs to a specialist group. One specialist group might carry out investigations on which materials provide the best insulation; another on ways of waterproofing materials; and another on stretchability of materials and how this affects the design of clothes. After the investigations the children return to their production team to share their knowledge. As each group member has carried out a different experiment they must concentrate during the investigation and ensure they understand the results so they can inform their companions. (See Figure 9.1a.) The composition of the specialist groups may be related to the children's attainment and the tasks given to them can take into account their ability.

It is possible to revisit the specialist groups several times to share and evaluate progress and to see how different groups are solving their problem with a view to returning to the production group with further ideas as shown in Figure 9.1b. This figure shows how specialist groups might each try out a different experiment which will help them to design a cart to travel up a slope and then share their knowledge with their production group who design and make the model.

The specialist groups could also carry out research from books or collect information on a visit. They could research different energy sources with the production group making a non-fiction booklet on energy. On another occasion, during a visit to an industry, specialist groups can spend time performing different tasks, such as interviewing various workers. These specialist groups can then share their findings with the production groups who compile reports on the visit.

This is a very effective approach that guarantees every child is involved. In the latter example, if the reports about the interviews were given to the whole class, only a few children would have the experience of the delivery or the reporting time might be excessive and beyond the attention span of the class. Each child knows they have a specific task for their group that is not duplicated by any other. The 'specialists' must make sure they understand and can express their ideas when they return to their own group, who need to listen to each 'specialist' to acquire their knowledge. This technique is particularly effective in involving the less able children, who can be helped by their peers in the specialist group to grasp the content and are highly motivated to do so. The production group, of course, will be motivated to encourage and assist their group members as they will need their knowledge and cannot lightly dismiss their contribution, giving the less able or reserved child value and confidence.

Discussion groups

In a discussion a group of children can work to share experiences, understanding, ideas and opinions on an issue in order to enhance individual understanding or to arrive at a group consensus. Young children might discuss

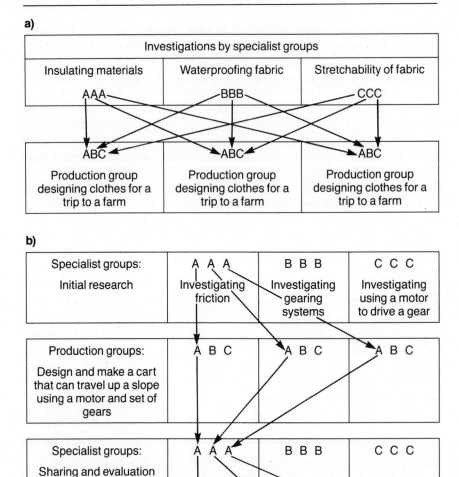

Figure 9.1 Organisation of specialist and production groups

how to solve problems raised by a story which has been read aloud to the class. For example in the story 'The Lighthouse Keeper's Catastrophe'[15] the keeper is locked out and when he goes to get his spare key it ends up at the bottom of the sea. Small groups can discuss ways of helping him get into the lighthouse or getting the key from the seabed. A list of possible solutions may be the only outcome of the group work, but it could lead into other discussions or a making activity such as producing their own versions of the story for a story-tape package; evaluating commercial tapes to identify what features they like and could include on their tape (e.g. signature tune at the beginning and sound effects); producing their story in book form with appropriate tape and book cover; or designing a key-ring that cannot be lost easily.

Older children might be given the task of discussing the difference between an advertisement and a factual account in a magazine, or what factors constitute an effective television advertisement. On another occasion they could be asked to give a three-minute talk on a choice of given topics, such as 'What is a technolog. What do they do?' or 'How can we help care for the environment?', which can require the children to research their subject in the library, discuss their ideas and clarify them in order to give their presentation.

Role-play activities and simulations

Role-play and simulation activities help children to develop cooperative group skills because they are able to suspend their normal complex class interrelationships and take on simplified characters within an event or situation. Very young children might role-play the participants in a shop, bank or café that they have already designed and set up. Older children could duplicate an industrial production line with each child in the group taking on one of the tasks in the line or they might take on the roles of industrialists, farmers, and employed and unemployed people in a public meeting where one of several new motorway or railway routes is to be decided upon. (See Chapters 3 and 7.)

DEVELOPING COOPERATIVE AND PROBLEM-SOLVING SKILLS

To fully develop cooperative and problem-solving skills, time and practice are required, and in order to motivate the children to succeed in their development these skills need to be seen to be valued by teachers. If the pupils do not receive the customary signs of 'proper work' such as feedback, marks and praise, the children do not rate such activities as work, and therefore tend not to take them seriously.

Developing children's collaboration

When children start working in pairs and small groups it is helpful to discuss rules they might like to draw up that would help them collaborate effectively. Initially the children tend to identify ways of social behaviour such as speaking quietly but so that everyone can hear; being polite to each other; listening to each other; encouraging quiet members to speak or participate;

sharing jobs and making sure everyone can participate fully. As the children become more mature and used to working together it is worth discussing what they might do, an addition, to assist the group to complete their project more efficiently. In this case they might come up with suggestions such as:

1. *Organising:* The group members think out how to share the tasks and in what order they need to be done or they may agree to take turns. There may be suggestions to increase or reduce the pace of the work, perhaps for more careful making or to encourage each other to meet a deadline.

2. *Supporting:* Weaker members are assisted. Others are praised for their efforts and encouragement is given to companions to attempt difficult tasks. Comments are made that reinforce or support the ideas of group members.

3. *Challenging*: When individuals question their peers, asking others to justify, elaborate or develop their ideas, the quality of the outcome is likely to be enhanced, as long as it is not done in an aggressive manner.

4. *Suggesting alternatives:* Several ideas are put forward rather than only one option being considered.

5. *Explaining:* Reasons are given for a possible action. Support for an idea is provided by suggesting an example or parallel idea.

6. *Commenting:* The child might think aloud in an open-minded way or wonder about something that raises the curiosity of others.

7. *Summarising:* Individuals could summarise progress or focus the others' attention on the project if they start to move off-task.

8. *Finishing:* The children recognise that their aim has been achieved. They may also refer back to the original criteria to check that all requirements have been satisfied.

The teacher may wish to focus on only one or two strategies at a time with the children, or use a similar list to help the children evaluate their own skills once a task has been completed. Feedback of the children's success in cooperation is essential. Some teachers put a tick on the blackboard by the group's name whenever they see positive instances of cooperation and share examples of the successful strategies at the end of the session.

Some children who are good at written work can be impatient or negative in a cooperative group, as previous success has been achieved through their individual ability. Direct criticism by the teacher of this is unlikely to be productive, as the intention is to motivate teams to work well together and dwelling on the weakness of a group member is unlikely to enhance the self-esteem of the whole group. Consequently it is usually more effective to praise others in their presence, so drawing their attention indirectly to their poor contribution in this situation.

Groups could be taped or videoed so that they can hear and see for themselves how they interacted. As this process is time-consuming different groups might be taped on separate occasions so that they can take turns in assessing their skills. As part of the final evaluation each child can try to assess their part in the group, and the whole group might assess what aspects made their group effective and what they could do to improve on another occasion.

Developing problem-solving skills

Children not only need strategies for working cooperatively but they also need guidance on ways to solve problems. After the children have had the experience of open-ended problem solving, the following different stages in the process could be identified with the intention of improving subsequent activities.

1. *Understanding the task:* The children need to put the instructions into their own words and to be clear about any limitations or specifications. Children often rush into an activity without looking at the whole problem, so have to make adjustments later. They might have been asked to produce a poster for nursery children warning of the dangers of playing with electrical plugs, and if they do not think of the intended audience it might be halfway through the making process before they realise that the language used in the poster will be inappropriate for the age of the children concerned.

2. *Review of existing knowledge that may be helpful and how similar problems have been solved by them or others in the past:* For example, if the children have been asked to find a way of helping the caretaker to lift a heavy sack of clay on to a table, they may be able to recall that lifting can be made easier by using slopes, pulleys and levers.

3. *Collection of different solutions:* The children need to be encouraged to put forward several solutions, even if some are not very practical ones, rather than rushing into the first method that comes to mind. Brain-storming to collect spontaneous expression of ideas without evaluation, criticism or discussion is a useful technique for collecting many ideas in a short time. To raise the sack of clay the children might suggest fixing a pulley to the ceiling, using a slope or even hiring a fork-lift truck.

4. *Ranking and choosing an approach:* The possible suggestions need to be considered in detail, perhaps with sketches to show how they might be achieved and equipment checked for availability. Bearing these factors in mind, and the time limit, one method is chosen. The children might draw a rough picture of how they could fix up a pulley in the classroom, either on its own or in association with a slope, and then check whether a long-enough flat piece of wood or actual pulleys are available or could be made with the materials that are to hand.

5. *Identifying the different tasks and deciding on an order for action:* This may be helped by drawing a flow chart.

6. *Making or carrying out the plan.*

7. *Monitoring and review:* Inevitably problems will arise which may be minor enough to be discussed and solved informally, but in a major project problems might need to be dealt with by going through the above procedure

again. When dealing with the sack of clay, for example, the children may discover that the material of the sack will tear under the weight of the contents as it is being lifted, so they will have to make adjustments to their procedure perhaps by reinforcing the sack or making some type of harness.

Children need time to build up patience and confidence to follow a procedure like this, as they enjoy the making process and will want to start it as soon as possible, but as they start to tackle more complex problems they will increasingly see the need for good planning.

Developing listening skills and negotiation

In cooperative groups children need to listen to each other, whereas they often only expect to take notice of the teacher. However, if the teacher regularly asks children to comment on another pupil's statement, they will start to value their peers' contributions. Teachers need to be careful not to repeat or summarise too often a pupil's contribution, which may have been spoken quietly or not expressed clearly, as this will nullify their intention to encourage the children to listen to their peers.

Paired oral work also provides a good foundation for listening to others and might include interviewing each other about a favourite item. Using this information, each pupil could explain their partner's choice to another pair of children. On another occasion pairs might have a short oral sharing of ideas which are then fed into a class discussion. For example, if the class is considering how to improve the environment of their school, pairs of children might make a list of their ideas which are then shared with the others.

Snowballing is an extension of this approach where a very small group of two or three children have to discuss a task in a limited time and then form a partnership with another group and compare ideas. This process is repeated with a larger group with a slightly different task at each stage to maintain the children's interest. Two children might be asked to brain-storm a list of ways of behaving well in a group. Then fours can be asked to share their lists and select six ideas. Finally eight children could combine their lists together and put them in order of importance to display on the wall. The advantage of this technique is that it encourages all the children to formulate their ideas and not just to rely on the dominant members of a large group, and it gives children the opportunity to learn to negotiate and come to a consensus.

Shared story-telling where a story is created by the children in turn, each having to take up the story from where the last left off, also requires children to listen very carefully in order to develop the actions and use the characters in the story. The children have to build on others' ideas rather than being able to produce a complete story individually, and this demonstrates how combining ideas can provide an exciting and imaginative group product.

Teacher intervention and children's independence

It is important both to develop the children's abilities in handling tools and materials, and to allow them to apply these skills independently in open-ended design problems. There is a view that teaching technological skills is a prerequisite so children have confidence and ability to reach a solution in problem-solving situations and that these should be taught separately from design activities. One of the effective ways of teaching skills is by giving children specific instructions to create a predetermined product with a high degree of teacher intervention. Once the pupils have a variety of skills the teacher may then feel it is appropriate to stand back completely so that the children can tackle a design problem without interference, being free to be more creative and so developing a sense of achievement and confidence in their own abilities. However this is too simplistic a division. When carrying out a controlled assignment, children can follow instructions rather automatically without understanding or appreciating why a skill needs to be performed in a particular way. Consequently it often helps to include elements of negotiation and discussion in these sessions, with the children suggesting how their skills can be applied and by the teacher varying the task slightly to respond to the children's comments. Total non-intervention during a problem-solving activity is unlikely to be satisfactory, either as the children's imagination is likely to raise the need for additional skills, without which they will be restricted and frustrated in achieving their desired outcome. Knowing the moment and method of intervention requires experience and skill on the part of the teacher.

The source of authority in the classroom is usually the teacher, but in cooperative open-ended activities it is hoped that the group will become the initial source of help. The children need to be told explicitly to ask for help from the group before approaching the teacher, otherwise they will request immediate intervention in minor disagreements or problems. Over time, as the children get used to this procedure, their demands on the teacher's time drops, and usually only requests that genuinely require assistance are made. Not only do the children gain in independence, but the overall quality of the language, thought and responsibility is also usually higher than in teacher-led groups.

Teacher intervention should not depend solely on requests from children. The teacher also needs to step in when the group is in difficulties or is pursuing an inappropriate course of action. The timing of this intervention is very significant, as children need to learn from their mistakes but must not become excessively disheartened. Sally Frost describes a child making a glove puppet who drew around his hand and cut out material without leaving a seam allowance.[16] Once he had tacked it together he discovered for himself that it no longer fitted his hand. He then drew around his hand again and cut out a larger version. It was not until he realised the problem that he could devise a strategy to solve it. If the teacher had told him to make a bigger pattern he may not have realised the significance of the advice. However the teacher might intervene to praise persistence and discuss what the child has learnt, or if the child appears to be becoming distressed could discuss the problem and prompt ideas for solving it.

In another school, groups of reception children attempted to make moving play equipment for a doll, weighted with sand to represent a real child. The children set about the task enthusiastically and with minimal assistance and produced some very basic models. On completion each one was tested with the doll, but the children found that the models were too small and weak and did not move. The nursery nurse handled the 'failure' very sensitively by encouraging the children to discuss their work and that of other groups in a positive way to say what was good and how it could be improved. The groups then used this 'prototype' to plan and make a very satisfactory series of models for a park that met all the criteria, although perhaps not in the anticipated way. The children learned a great deal from their experiences, not the least that failure can be a very positive event if used as a learning experience and the freedom to be independent enabled them to be more imaginative than the teacher's preconceived ideas.[17]

Primary schools traditionally create an unreal world where failure and unpleasantness are avoided. Problems are frustrating and disappointing and it is important to acknowledge these feelings, but they do not have to be regarded as failures. Facing a problem and finding a way around it, perhaps with adult advice, is far more worthy of praise than a solution that conforms with the teacher's preconceived ideas and that was produced without much effort or thought. Indeed, the 1990 National Curriculum Design and Technology indicates that teachers should assess whether children can 'adopt alternative ways of carrying forward their plan when difficulties are encountered and recognise when help is needed'.[18]

The approach of the teacher is very significant. In technology children need to face problems but they also should know that help is available; that others will not be allowed to criticize unkindly; and that the teacher will intervene before mistakes are irretrievable. In order to be able to provide advice teachers need to know how to solve a set problem, but they will usually find that even very young children, or those identified to have special educational needs, are able to surprise them with the variety and imaginative approaches to problems, once awareness that the teachers want children to be independent is established. Consequently teachers often learn as much as, if not more than, the children. For many teachers giving children greater independence requires a shift in their teaching approach and it needs to be recognised that teaching styles for design and technology are a great deal more demanding than the comparative order and control of 'traditional' styles of teaching. However, when it works, 'standing back and watching groups of highly motivated children using their own initiative and resources to extend their experiences and capabilities significantly is an experience that few teachers forget in a hurry.'[19]

Developing leadership skills

The teacher must decide whether to give a specific child a leadership role or to let a leader develop naturally within the group if they require one. A leader is not always necessary for the effective functioning of a group. Biott[20] found many examples of different pupils within the same group raising the quality of the group's work in ways which often surprised the teachers, who had not witnessed such behaviour previously.

Sometimes the teacher may decide to make an 'official' leader so that different children can take a turn to be explicitly given the opportunity to develop leadership skills. The teacher can then support the children in that role by referring all requests through them and deferring to them. The role of the leader could also be discussed with the class, who can try to identify what special tasks leaders should do, such as helping the group get organised, being the final decider of the allocation of tasks, liaising between teacher and group; protecting the interests of all the children in the group; practising good cooperative behaviour; and helping with disagreements by impartiality. If the children have discussed how a leader should behave, this might avoid the problem of 'official' leaders becoming excessively bossy and impatient with the group or being limited to only reading instructions, writing or delivering the report, whereas their role can be very different and actually more demanding.

There are problems with children working in unsupervised groups, such as undervaluing some members; the dominance of a few by talking too much or assigning low-status tasks to their peers and keeping all the high-status tasks for themselves; the breaking up of the group into factions; failure of some children to take an active part; or an acceptance of an over-easy solution to avoid facing complex issues. Many of these can be avoided or at least greatly reduced by choosing the group size and composition carefully and discussing how to work effectively in a group beforehand. A very powerful influence is the way praise is used by the teacher. By identifying good models of group behaviour and commenting on it, the teacher avoids humiliating the child who finds cooperative behaviour difficult but reminds the children of what is a good approach and demonstrates that it is highly valued by the teacher. It is also important that the teacher acknowledges that sometimes others in a group can be exasperating, and compromise is difficult for everyone, even adults. The evaluative process specifically required by technology also enables the children to be asked to review their own behaviour with a view to identifying how it might be better.

NOTES AND REFERENCES

1 National Curriculum Council (1990) *Non-Statutory Guidance: Design and Technology* York: NCC, B7 Section 2.12.
2 Galton, M., Simon, B. and Croll, P. (1980) *Inside the Primary Classroom* London: Routledge & Kegan Paul.

3 Bennett, N., Desforges, C., Cockburn, A. and Wilkinson, B. (1984) *The Quality of Pupil Learning Experiences* London: Lawrence Erlbaum Associates.
4 Dunne, E. and Bennett, N. (1990) *Talking and Learning in Groups* Basingstoke and London: Macmillan.
5 Cowie, H. and Rudduck, J. (1988) *Co-operative Group Work: An Overview* Sheffield: BP Educational Service.
6 Aronson, E., Blaney, N., Stephan, C., Sikes, J. and Snapp, M. (1978) *The Jigsaw Classroom* London: Sage; Johnson, D., Maruyama, G., Johnson, R., Nelson, D. and Skoa, L. (1981) 'Effects of co-operative, competitive and individualistic goal structures in achievement: A meta analysis' *Psychological Bulletin* Vol. 89, pp. 47–62; and Yeomans, A. (1983) 'Collaborative group work in primary and secondary schools: Britain and the USA' *Durham and Newcastle Research Review* No. 51, pp. 99–105.
7 Johnson, D. *et al.* (1981), op. cit.
8 Alexander, R., Rose, J. and Woodhead, C. (1992) *Curriculum Organisation and Classroom Practice in Primary Schools* London: DES.
9 Morgan, V. (1989) Primary Science – Gender differences in pupil responses *Education 3–13* Vol. 17, No. 2, pp. 33–7.
10 Smail, B. (1984) *Girl-Friendly Science: Avoiding Sex Bias in the Curriculum* York: Longman; and Kelly, A. (ed.) (1987) *Science for Girls?* Milton Keynes: Open University Press.
11 Tann, S. (1981) 'Grouping and group work' in Simon, B. and Willcocks, J. (eds) *Research and Practice in the Primary Classroom* London: Routledge & Kegan Paul.
12 Bennett, N. and Cass, A. (1989) 'The effects of group composition on group interactive processes and pupil understanding' *British Educational Research Journal* Vol. 15, No. 1, pp. 19–32.
13 Ibid.
14 Howe, C. (1990) 'Grouping children for effective learning in science' *Primary Science Review* No. 13, pp. 26–7.
15 Armitage, R. and D. (1988) *The Lighthouse Keeper's Catastrophe* Harmondsworth: Penguin.
16 Frost, S. (1990) Children solving problems? in Tickle, L. (ed.) *Design and Technology in Primary School Classrooms* Lewes: Falmer Press.
17 Jarvis, T. (1989) 'Toys that move' *Child Education* October, pp. 39–40.
18 DES and Welsh Office (1990) *Technology in the National Curriculum*, Attainment target 3: Level 4d London: HMSO.
19 Shepard, T. (1990) *Education by Design: A Guide to Technology Across the Curriculum* Cheltenham: Stanley Thornes.
20 Biott, C. (1987) 'Co-operative group work: pupils' and teachers' membership and participation' *Curriculum* Vol. 8, No. 2, pp. 5–14.

FURTHER READING

Ainscow, M. and Muncey, J. (1989) *Meeting Individual Needs in the Primary School* London: David Fulton.
Biott, C. (1984) *Getting on without the Teacher: Primary School Pupils in Co-operative Groups* Sunderland: Sunderland Polytechnic.
Burden, M., Emsley, M. and Constable, H. (1988) 'Encouraging progress in collaborative group-work' *Education 3–13* Vol. 16, No. 1, pp. 51–6.
Fisher, R. (ed.) (1987) *Problem Solving in Primary Schools* Oxford: Basil Blackwell.
Galton, M. and Williamson, J. (1992) *Groupwork in the Primary Classroom* London: Routledge.

Holt, J. (1984) *How Children Fail* revised edition, Harmondsworth: Penguin.

Ruff, P. and Noon, S. (1991) *Signs of Design: English* London: Design Council.

Simon, B. and Willcocks, J. (1981) *Research and Practice in the Primary Classroom* London: Routledge & Kegan Paul.

Tickle, L. (ed.) (1990) *Design and Technology in Primary School Classrooms* Lewes: Falmer Press.

Williams, P. (1990) *Teaching Craft, Design and Technology: Five to Thirteen* London: Routledge.

Chapter 10

Assessment

Assessment serves several related but different purposes. Primarily it is concerned with the evaluation of pupils' progress and achievement as a formative, diagnostic measure of particular strengths and weaknesses so that appropriate activities can be provided. It should also enable teachers to evaluate their own performance and to assess the effectiveness of the content and approach adopted by the school. In addition, more formal summative assessment is required to identify levels of achievement so that parents, governors and outside bodies can be better informed, hopefully in order to provide suitable support, assistance and resources for teachers and pupils. It is important that assessment does not control the teaching approach and type of experiences offered to the children, which should be chosen primarily in respect to their interests rather than to satisfy the requirements of an external agency, although in many cases a method can be adopted to satisfy most needs. Assessment 'should be the servant and not the master of the curriculum. Yet it should not be a bolt-on addition at the end. Rather it should be an integral part of the education process, continually providing both "feedback and feedforward".'[1]

Although optional Standard Assessment Tasks, as models of ways of assessing design and technology, are available for schools to use, most assessment throughout the primary school will rely on teacher's professional judgement based on continual monitoring of children's progress. In addition, day-by-day teacher assessment will also have to include skills and concepts such as understanding systems and ideas of energy, control and structure outlined in the Programmes of Study which are not included in the attainment targets.

The National Curriculum Council appreciates that 'the means of assessment in design and technology may need to be different in some respects from that in core subjects because there should be a greater emphasis on process than content', and, in addition, as 'each attainment target relates to the others and cannot be treated as a separate stage of a linear process'[2] the assessment of design and technology should be based on the whole task. Anything less than this, for example a test of pupils' understanding or skill independent of such a context, would lose validity[3] and would be contrary to the nature of designing. The significance of process and skills, and the fact that they should not be assessed as isolated formal short 'test' activities, makes assessment in design and technology particularly difficult.

Many elements of design and technology – including cognitive understanding of ideas such as energy transfer; manipulative skills and the appropriate use of tools and materials; cooperative skills, planning, organisational and evaluative abilities; and imagination and aesthetic awareness – do not easily lend themselves to objective assessment. Good ideas for solving a problem are not always a result of a conscious, analytical and structured process but can be a flash of inspiration. Evaluating the aesthetic element of a product is subject to very personal views, and as much of design and technology will be carried out in groups, in order to develop the essential skills of leadership, teamwork and cooperation, there is a problem in identifying the individual's level of participation and success.

Relying only on looking at the final product for assessment purposes will be inadequate as there are no right or wrong answers, and a successful product from a particular assignment may be achieved in different ways and take many forms. For example a well-made functional item may appear to be better than one which shows equal skills but does not meet all the original specifications. However, the first children may not have successively achieved as many skills as the second group during the design process, who might have shown a greater ability to allocate and share tasks, consider procedures to minimise waste, and were able to identify how their product could have been improved. Much assessment, therefore, will rely on observation during the teaching process. Indeed during practical activities children are producing evidence of their skills and understanding almost incessantly and simultaneously. Unfortunately many signs of this attainment are ephemeral, such as a few spoken words commenting on the value of a procedure or a decision not to take a particular course of action because of the religious feelings of the customer. These incidents may demonstrate unexpected skill and understanding but are difficult to collect as evidence. In addition to these problems, as teachers are unfamiliar with teaching technology they are also inexperienced in recognising when a child has achieved a particular skill.

Despite the fact that assessment in design and technology is difficult it is essential, otherwise there is a real risk that children will not be presented with suitable tasks that relate to their development and will ensure a balanced experience across the very broad range of the subject. Much of this assessment will have to be through observation during the teaching process. In order to do this effectively, teachers need to integrate planning, teaching and assessment; to build up skills of observing and analysing technological activities; and to acquire experience of how different children respond to these activities. This assessment process can be assisted by:

1. *A whole-school policy indicating when different skills and concepts will*

be introduced: The Technology National Curriculum gives teachers a framework for their planning and indicates what skills and concepts are appropriate at different Key Stages, although it does need further detail. (See Chapter 2.) Such a planned progression helps teachers to know what to focus on in their assessment and what the children should already know. Additionally, if the whole school plan is produced in collaboration with all the teachers, they should come to a clearer, shared understanding of what children can do and how this might be recognised.

2. *Pre-planning assessment for individual sessions:* By pre-planning assessment, teachers are more likely to observe the children in a systematic way with less risk to preconceptions affecting their judgements. Such planning will also have the effect of increasing their skills in understanding the Technology National Curriculum. Teachers will need to consider both the attainment targets and the aspects of the programmes of study that will be covered by a particular activity as, unlike many other subjects, they vary significantly. Initially, as teachers are building up their experience of assessing technology, they might concentrate on assessing a limited number of children at a time.

Sometimes children do not respond as planned. They may do more than the task demanded or something completely outside that which the teacher expected. In this case the teacher may wish to note down significant learning demonstrated after the activity. It will not matter if the planned task does not progress as expected or proves to be inappropriate for the children and has to be abandoned, as long as assessment is a frequent element of all teaching in order to build up a changing picture of the children's improvement.

3. *Making notes or records from observations during the design process:* Observations of individuals and groups may be carried out by the teacher during the activity or later by reviewing material collected on tape recorders or videos. When assessing individual attainment within a group situation the teacher needs to note how the tasks within the group have been shared, the level of individual involvement in the work, how much time is spent on-task, and what strategies the children have used to help each other to reach a satisfactory outcome.

4. *Consideration of the final product:* It can be expected that the products will give an indication of the skills applied in the making stages, although observation and discussion with individuals and groups will be needed to get more detail and to ascertain what contribution was made by individuals in group contexts.

5. *Encouraging children to keep working notes, plans and diagrams:* The teacher should ask the children to keep any rough sketches they produce in a folder. Children tend to throw these away as they are usually drawn on scraps of paper or are only partial drawings. However, if these are obviously valued, the children are more likely to get into the habit of keeping them and recording more of their thought processes. These drawings can be used as

evidence of planning in their own right or as prompters to help the children to recap their activities so that they can write about or discuss the work and individuals' contributions in the group later.

6. *Children's self-evaluation and overall evaluation of the design activity:* 'Pupils should as far as possible be involved in the assessment of their own work'[4] and this can be more effectively achieved if the objectives for the sessions are shared with them. In addition such self-assessment is particularly valuable in design and technology as it helps to identify the level of individual performance in group situations. The evaluative phase of the design process, when groups write about their project, produce a display or tape recording, or give an oral report to the rest of the class, will also provide opportunities for the teacher to ascertain the children's level of ability. However it must be remembered that if information is omitted it does not mean that the child does not know it. This can only be discovered by further probing and questioning.

7. *Giving assessment tasks that focus on specific skills or concepts:* These might include checking that each child can use a glue gun or sewing machine correctly, or using drawings and concept mapping to determine a child's level of understanding of energy sources.

METHODS OF COLLECTING EVIDENCE

It can be seen that assessment of individuals working alone and in collaborative situations requires a range of assessment methods. Consequently teachers need to be aware of the possibilities and their advantages and limitations, so that they can choose the most appropriate procedures with regard to the type of activity, age and ability of the children and their own expertise in teaching and assessing technology.

Teacher observation

Observation is one of the major ways of collecting evidence for children's attainment in design and technology. There are three main types of observation: open, non-specific; focused; and systematic. In the first, teachers look at the activity and children's responses and note down points that capture their attention. There are considerable limitations in this approach, not the least that the teacher is fairly easily influenced by their expectations and by the dominant behaviour of some of the children in the classroom. The TGAT report comments on the 'halo effect' where, in the absence of a close definition of what to look for, teachers see confirmation of their expectations. A child who is judged to do well in one area is given unmerited favourable assessments in other areas, or the opposite where a 'labelled' child is not credited with an achievement because it was unexpected. This can be avoided to a certain extent by having pre-planned focused observations, whether of a particular child, group, specific skill or application of knowledge.

Systematic observations are planned in very great detail with rigorously defined categories and often involve recording behaviour on time intervals. Although this technique produces data that is very explicit and lends itself to producing data in quantifiable form, it is very time-consuming and can only be done effectively as a separate activity from the teaching process.

Focused observations therefore will usually be the most appropriate for the classroom teacher as a method of observation that is relatively objective and can be managed as part of the teaching process. It has the advantage that it is flexible and need not interfere with normal activities or take up excessive preparation and recording time. It can be used frequently and provides constant feedback to enable work to be matched more closely to the needs of the children. Special equipment is not required and the children do not necessarily have to be aware that they are being assessed.

Before each technology activity the teacher should identify the main skills, concepts or areas of knowledge to be covered. The clearer the aims of the activity are identified the more likely they will be both achieved and observed. Rather than to attempt to assess all skills and concepts, two or three only might be chosen. For example the project may be to design and make a container to carry something fragile such as a pet carrier, lunch box or soft fruit punnet. Although the children will meet all the attainment targets during the project, the teacher may choose only to assess whether the children are able to use paper or card to make three-dimensional prototypes to form the basis for accurately measuring a pattern onto corriflute (plastic corrugated sheet); whether Stanley knives are used correctly and safely; and if the children use a procedure that minimises waste.

Teachers must not feel that they should be expected to observe every child and note every skill or concept achieved in one session. However records need to be made to ensure all children and all skills and concepts are observed over time. It is also important to observe the same child on several occasions to ensure that one poor performance was not the result of the child being tired or ill, being unwilling to make a personal commitment to the activity, or being unduly influenced by others in the group. The repetition of observations is particularly important in design and technology because the projects will vary enormously from sewing drapes, producing a dance, designing a garden, to making a working engine, and are likely to have very varied appeal to different children. As the children will also be working in groups, repeat observations are needed because the interaction and ideas suggested by members of the group will influence the performance of the others.

Observation could be carried out as part of the normal teaching, or another approach is to have short periods, or five or ten minutes, when observation is the teachers' prime aim. The children will need to understand that they should not interrupt the teacher during these periods. Usually the children respond very well to this situation, particularly if they are informed that this activity is to help the teacher learn more. A notice or symbol, perhaps drawn on the board, to remind the children that the teacher is not to be disturbed can be useful and the children need to know what do to if they are stuck during periods of observation. Indeed this activity can also be used to help the children develop greater independence, which will assist not only in their approach to design and technological work but also in creating opportunities for observation and assessment alongside normal teaching.

The children's autonomy can be increased by ensuring that the materials they need are organised so that they are easily available and by making sure that the children know how to use them. The class might also discuss procedures for self-help in order to draw up a list of prompters, such as: Is there a book or a picture that will help? Have I solved a problem like this before? Can I get the material I need? Am I allowed to use the tool by myself? Can anyone in my group help? In the latter case children need to know that collaboration is acceptable. It is often useful to discuss with the children the difference between telling a friend the answer outright and guiding them to understand or finding the answer for themselves.

While watching children for assessment purposes, teachers should avoid premature or inaccurate interpretation or being tempted to interrupt in order to assist or further the children's learning, as this may change the direction of inquiry or stop a child showing competence in a skill. The teacher will almost certainly identify the potential of a line of inquiry or see the solution to a problem before the children. However, given a little time, the children may come to the same realisation for themselves. If a particular child is the chosen focus for assessment for that day it might be better not to intercede at all, but if the teachers feel intervention is in the best interests of all the children they should endeavour to pose open questions which leave the children the choice of action.

Teachers need to build up experience in observing design and technology by initially observing only one or two children and by occasionally concentrating on observation rather than giving teaching and observation equal emphasis. As part of in-service training it would be helpful if colleagues could occasionally team-teach so that the teachers could take turns to concentrate on observation. In addition to building up their expertise and confidence by taking a step-by-step approach to assessment, teachers might also match their assessment against self-assessment by the children and by occasionally using videos of children working. The latter enables assessments to be made away from the immediate pressure of teaching and allows groups of teachers to discuss their assessments of the same incidents. One teacher who took a video of groups working on technological activities was amazed to discover that, although in many cases his observations were accurate, on others he had missed significant incidences because of his necessary involvement in another part of the classroom. On one occasion he felt that one child had made no effort at all to interact with an activity but, on watching a video later, discovered that the boy in question had actually given all the imaginative ideas to the group and then quarrelled with someone and subsequently taken no further part in the proceedings. On another occasion a group, which had produced a very imaginative product showing application of advanced concepts, were seen on video to have slavishly followed the instructions of an adult helper. The teacher involved felt the experience demonstrated the value of using a variety of assessment techniques, including the children's self-assessment in particular.

Pupils' self-assessment

Design and technology lends itself to developing pupil self-assessment because evaluation of the planning, making and end product are required elements of the subject. As part of evaluation the children might try to suggest how much they have learned and what part they as individuals have made to the final product. This might be achieved as a whole-group exercise producing a collaborative explanation of how the group worked with each other and what part each child made. On the other hand self-assessment could be carried out as an individual activity, either orally with the teacher, as a simple questionnaire or as a written report. The children could even design the proformas as a technological project, including such questions as:

- What I liked/disliked about the session.
- What I found easiest/hardest.
- What I think I have learnt.
- What I found difficult to understand.
- Ideas or things that I did that I am particularly pleased with.
- How well the group worked together.
- How did working together help.
- How I helped the group to complete the task.
- How other people help me.

In addition to these general points the children might usefully be given the main aims of the session and what skills and knowledge the teacher hoped would be covered, and asked to comment on these in particular. (See Figure 10.1.) This should also help them to appreciate what the teacher values and is trying to teach, and often enhances the children's desire to achieve these aims. Given a supportive environment, even very young children are very honest and remarkably accurate in their judgement.

Examples of work chosen by the children could be kept in a folder to build up a profile of the children's development. These could include diagrams, written reports, photographs of models and tape recordings of presentations, with the children's comments on why these particular items were chosen.

Peer evaluation of products also enables the teacher to make assessments. As part of the evaluation process each child or group might display their end product and explain the planning process, successes and problems met and

Concepts/skills aimed at by the teacher	Activities done	How well did you do? 
Ability to link properties of materials to use	Drawing and naming parts of the bicycle	
	Labelling materials used on a picture of a bicycle	
	Reasons why the manufacturer used each material	
Appreciation that a triangle gives strength to structures and how this idea is applied in bicycles	Testing different shapes to find the strongest shape	
	Inventing your own strong shape	
	Finding strong shapes on the bicycle	
Setting up fair tests	Setting up a fair test to find out whether a flat piece of paper is stronger or weaker than a tube	
Appreciation that materials can be strengthened by changing their shape – a tube in this instance	Finding out which shape tube is strongest	
	Finding how tubes are used in bicycles and around the school	

An additional column can be added for pupils' comments in which they are encouraged to comment on their particular successes and difficulties.

Figure 10.1 Pupil self-assessment: The design of bicycles

how difficulties were overcome. The rest of the class could be encouraged to ask questions to identify what is good about the product and what ideas they might find useful on another occasion. Questions posed by other children may well raise significant aspects not considered by the teacher, as the children are aware of similar problems that they also met in the activity. By emphasising positive learning and ideas for improving future tasks, it is possible to encourage such peer review without damaging individual-pupil self-esteem.

Considering the product

Plans, sketches and diagrams of the planning process and the final product will help the teacher to place the child's work at an appropriate level. When individual end products are required by the task, even if these are created while the child participates in a group, each child can be fairly easily assessed. The main problems are that the pupil may have copied others or have relied on others' ideas without producing his or her own. These situations might be identified during on-going observation or by individual and or/group evaluation.

When one product is required of a group, individual contribution is far more difficult to ascertain. In addition to observation and self-assessment, the teacher might consider post-task written reports, individual interviews or whole-class discussions which require the child to demonstrate what they have learned.

Written reports, tests and summaries

Written reports and tests can be integral to the task or a separate assessment activity and might include multiple-choice questions, completing provided sentences, answering specific questions or extended writing. Such tasks are often seen as the usual and most effective methods of assessment but they have considerable limitations when assessing design and technology activities. They are very useful to ascertain children's level of knowledge but are far less suitable for assessing skills and concepts which are particularly significant in this subject. Such techniques also presume that the children have an adequate level of reading and writing ability and are able and prepared to express themselves about their design experiences in a written form. If this is not the case, children who might otherwise be enthused by and excel in technology are discouraged. It is important that children can demonstrate their technological abilities uninhibited by problems of language competence.

Multiple-choice questions can be used by children with a wide range of reading ability as the questions can be read out to them or prepared on tape, and the answers do not need any detailed writing. Such questions are very quick to mark but are unfortunately time-consuming to prepare, and should include common misunderstandings in the choices, requiring the teacher to have some depth of knowledge of the subject. On the whole this technique is only of any value for testing knowledge.

Questions, cloze passages or sentences to complete are easier to prepare than multiple-choice questions and manageable for children with limited writing ability. They are suitable for testing knowledge and comprehension but are of little value for assessing skills, and in order to show a grasp of concepts the children probably require long answers using unfamiliar technical language which may make written expression difficult for them. The provision of a partially completed passage or sentence may assist some children, whereas others can be inhibited as they struggle to find the answer they feel is required by the teacher. Such approaches are very valuable parts of technological activities as a means to help children to clarify their ideas and to use new vocabulary, but it should be realised that they have limitations for assessment.

Extended pieces of writing such as 'Explain how you arranged a bank account for your mini-business' allow children scope to write about their knowledge and will probably demonstrate their level of concepts. Children who are reluctant to write, or find the process difficult, might usefully prepare an oral report on tape instead. Children may give a great deal of detail on one aspect and not cover another, or they may go into considerable detail about an incident that was significant to them but do not cover those aspects the teacher wanted to assess. For example a child may report in great detail about the welcome and relationships between the people in the bank and say very little about the paying-in mechanisms, use of cheque book and statements, and what the bank did with customers' money that were the main foci for the visit. It might be helpful, therefore, if the teacher suggests subheadings. Even so it cannot be assumed that a child does not have grasp of a concept because it is not included in their report.

Diagrams and drawings

Unaided drawings and annotated diagrams produced during activities as well as those drawn as a summary can be useful for assessment. The children might be asked to draw or arrange pictures of people to represent a hierarchy after a visit to a factory, or to annotate the forces that are affecting the structure of a model bridge as the summary of a project on the design of bridges. Such drawings can also be very valuable ways of assessing children's understanding before a topic is started and as a stimulus for class discussion. A teacher, having decided to study shops with an early-years class, could ask the children to draw pictures of what happens to the money that is placed in the shopkeeper's till or ask the children to draw a topic web centred around the word 'money'. The pictures will give an idea of general class understanding as well as indicate individuals with a very advanced understanding and who is likely to need additional support. The teacher is then able to plan the details of the project and groupings with this information in mind.

Concept mapping

The concept map is a schematic representation of relationships between concepts and again can be used for assessment both before and after teaching a topic. The starting point is usually a list of concept words, or photographs and pictures in the case of very young children and those with limited reading ability, that are known to the children and which can be sensibly linked together. A class about to study a local food-packaging firm might be given words such as 'food-packaging factory, farmer, shopper, shopkeeper, bank, delivery van, farm, supermarket, market and sawmill'. The children are then asked to write the words anywhere they like on a page, draw lines between the words and write 'joining' words on the lines. An arrow might be drawn from 'sawmill' to 'food-packaging factory' with 'wood bought to make boxes' written on it. Some words will be joined by several lines and other words will not be linked at all, giving an indication of the children's understanding of interrelationships. Assessment is greatly enhanced if the children also talk about their concept map. In addition to its use as an assessment tool for individuals, this activity is very valuable as a focus for group discussion which enables children to articulate, compare and clarify their understanding.

Written tasks and drawings can be useful for assessing knowledge and sometimes concepts but they are rarely of any value in assessing level of skills. This can only be consistently achieved through observation, discussion and questioning.

Discussions and interviews with individual children

Discussions and interviews give opportunities to gather information about skills, concepts, knowledge, feelings, attitudes, actions and reasons behind them. A major limitation of this technique is that it is time-consuming, although several children might be interviewed at once, unless there is a risk that one child might be ridiculed by others. The interviews can range from being highly structured to unstructured, or a combination which starts with a series of clearly defined questions and then on the basis of the answers probes and explores further in a less organised way. There are three different types of interview that can be used in assessing design and technology:

- A structured or standardised interview in which a predetermined set of questions is asked in a fixed order.
- A focused interview where attention is directed at a particular topic or theme is the most usual questioning method of assessing children's development in design and technology. Even in focused interviews a good deal of preparatory thought is needed to find key questions which will in sequence lead the children to express their knowledge and understanding.
- A conversational interview in a relaxed situation in the classroom or

playground as part of on-going activities where a particular issue crops up. In these situations children can offer telling insights that might not have been disclosed at another time.

In order to make maximum use of the interview children need to feel they can express their ideas fully without criticism, otherwise they will concentrate on producing what they think the teacher wants to hear rather than what they believe. The teacher therefore needs to be sympathetic, interested and attentive, without taking too active a role and without expressing an opinion. The language used should be appropriate with examples based on familiar contexts with, ideally, bilingual children having the opportunity to talk to an adult who speaks their home language fluently. Children also need time to think out their answers, as a 'don't know' may be a reply to gain time to gather thoughts rather than an indication of lack of knowledge. Some children might not be prepared to respond in a frank and honest way to a teacher but may be prepared to interview each other, particularly if the teacher prepares an interview schedule for the pupil interviewer. It is useful to tape these interviews for future reference, particularly if the sessions are relatively short.

The actual questions posed are important. Closed questions which require either 'yes' or 'no', or single correct answers where only one answer is acceptable are unlikely to be of much value except where they are used to relax the child and give them confidence. Open questions that allow for opinion, speculation, hypothesising and development of an argument are of far greater value. For example the question 'What tasks does the shopkeeper do?' gives children a greater opportunity to demonstrate their understanding than 'Did the shopkeeper pack the food in tins before selling it?', or 'How do you choose which saw to use?' should provide a far better appreciation of the child's skills than 'Should you use a tenon-saw to do this particular task?'

Ascertaining the level of children's concepts is difficult. One day a child may seem to grasp an idea but the next it appears to have been forgotten, as concepts established in one context may not be transferred to others. A series of questions increasing the application of experiences may indicate the level of understanding. Knowledge could be pictured as a pyramid. At the base of the pyramid are those aspects of knowledge characterised by information, data or facts. At this level the children have noticed that the shopkeeper sells one particular item for more than it was bought for. Concepts are developed in the second level when the children have identified simple patterns, groupings, reasons and explanations based on the facts. For example, when the children have accumulated a number of experiences that indicate that virtually all goods are priced at a higher amount than the buying price, they have grasped a basic concept. The third and highest level is where groups of concepts link to give generalisations or principles. At this level children will appreciate that profit is determined by many factors including buying price, interest rates, market demand, costs of premises and so on. Research has shown that teachers' questions are predominantly concerned with simple data and the recall of facts already learned.[5] Although teachers need to be sure pupils have a basis of facts before they progress to speculation and generalisation, they will only discover if the children have grasped the concepts behind these facts by asking higher-order questions.

After studying a food shop the children might be asked questions about the activities of the shopkeeper they met to see how much they had understood about their actual experience. A further stage could then be to ask the children how the shop and the shopkeeper's tasks might have differed if the shop had been selling something else, such as shoes; or what differences might have been seen if the visit had been in the summer instead of during the winter; or what would they advise the shopkeeper to do if faced with a particular problem such as food perishing. These answers should indicate whether the children are thinking logically about the experiences. Further questions could follow to ascertain whether the children could make generalisations. They could be asked, 'What things do shopkeepers need to think about when they stock their shops?'

Whole class or group discussions

In a class or group discussion, although the time taken is less than for individual interviews, it is more difficult to ascertain each child's attainment and to record findings before they are forgotten. After group work the whole class might come together to explain their work. If a quiet or low-ability child is asked to make the report it is a valuable way of assessing their understanding and therefore the likely understanding of the others. The other children can be asked to add their comments as a superficial check.

In question-and-answer sessions some pupils will want to answer everything, whereas others will appear to want to respond to nothing, perhaps because they are shy, unable to understand or are afraid of ridicule. In order to maximise the opportunities of assessment, teachers will wish to encourage answers from the less responsive, who often try to sit where they think they will not be noticed. Most teachers have a zone of maximum interaction, having a left-hand or right-hand bias or a type of tunnel vision. By being aware of this tendency teachers can deliberately look in those areas they do not naturally scan frequently. There is also research that indicates that teachers unconsciously respond more to boys than girls.[6]

Teachers not only tend to ask too many questions, they often ask them too rapidly. Failing to pause after a question leads children to give short replies and ones that call for memory recall rather than high-level thinking. Able children will dominate as the slower children cannot formulate their answers quickly enough, and therefore do not participate. Researchers indicate that if teachers train themselves to increase the waiting time, the length of pupil response increases; the number of unsolicited appropriate replies goes up;

pupil confidence increases as does the number of speculative answers with more thorough and evidence-inference statements; the children themselves start to ask more questions; the less able children make more contributions; and student-to-student interaction increases.[7] Teachers might take a tape recording of a class discussion as a personal check to assess whether they ask open questions, give children enough time to reply, and are unbiased in their approach to asking for contributions, and whether the level of language they used is accessible to all children in the class.

Use of photographs, tapes and videos

Photographs taken during activities can be used in individual and class discussions as an aid to recap the processes and skills undertaken. The children can be asked to comment on what they were doing and on what they thought others were doing. Such activity is helpful in focusing the children's comments and reminds them of their actions, and can be very helpful in revealing knowledge, understanding and reasons behind the application of particular skills. However films and processing are expensive and there is a delay waiting for processing. They are purely visual and are rather difficult to be seen by the whole class at once.

On the other hand, once bought, tape recorders are cheap to use and can be placed in different parts of the classroom where they are relatively unobtrusive. However replaying and transcription is very time-consuming and can only be used occasionally, given primary teachers' constant pressure of time. Unless the voices of the children are well known it is also often difficult to separate comments out and attribute them correctly to individuals, and some comments will be lost because of background noise or because a particular child has temporarily moved away from the group. However tapes do allow repeated analysis and provide physical examples of evidence which might be valuable to support more ephemeral observations. The tape might also usefully be used to help children to evaluate and assess their own performance, particularly with regard to developing cooperative group skills. In the same way a video recording can be used for both teaching and assessment. These recordings have the advantage of both sound and vision, but are expensive to purchase and a camera is difficult to make unobtrusive, although with repeated use children learn to ignore its presence.

USE OF RECORDS

Records and notes made regularly are essential to ensure that all the children are observed on several occasions. Repeated assessments are necessary, especially in technology, because observations cannot be completely objective, and the teacher may miss what a child is doing in the rush of the class, over-concentrate on the disruptive demanding children, or not observe the quiet, retiring individuals. In addition, inferences can only be made over time as factors such as motivation, behaviour of other children and physical fitness could radically alter the teacher's perception of each child on a particular day. Records are also needed to ensure that the whole range of technological skills are covered. Without a record system the teacher may observe the same aspects or skills frequently, but neglect others.

These records may take several forms, from a teacher's personal notes on individuals to a formal collection of work completed, with comments by the teachers, pupils and sometimes parents. Probably most teachers will use several methods. They might initially produce a tick sheet related to the particular project being undertaken, which is then filled in with reference to various assessment methods and pupil self-assessment. Notes might also be kept on individuals and samples of work saved. This information can be used later to complete summary sheets relating to the need to produce National Curriculum teacher assessments.

Recording achievement should not dominate the teacher's life. The aim therefore is to produce a system that is simple to complete, but includes all relevant information so that future action can be readily seen, be meaningful to others so that it can be used for communicating and reporting, and be accessible to pupils to enhance their understanding about their progress.[8] Consequently, rather than have a special record system for technology, it would seem more appropriate for it to be in accordance with the overall approach taken by the school which will be more readily familiar, making completion and communication easier.

CONCLUSION

Experienced teachers are able to glance at a child's English and mathematical work and have a good idea of what the child understands, where he or she is struggling, and what needs to be taught next. This expertise does not develop overnight and can only be acquired by accumulated experience with children. As teachers become more skilled in observing and assessing technology, they should find that their feeling for the subject increases and that they start to develop a mental picture of the progression of the design and technology curriculum and how children demonstrate their achievement in these stages in the same way as they have done with mathematics and English in the past. Teachers should expect to need this time to develop their skills of assessment and technology. If they set out to educate themselves using a step-by-step approach, they are likely to achieve equivalent competence fairly rapidly.

NOTES AND REFERENCES

1 DES (1988) *Task Group on Assessment and Testing: A Report* London: HMSO, para. 4.

2   National Curriculum Council (1989) *Technology 5–16 in the National Curriculum* York: NCC, Sections 6.1 and 4.1.
3   DES (1988) *National Curriculum, Design and Technology for ages 5 to 16* London: HMSO.
4   Alexander, R., Rose, J. and Woodhead, C. (1992) *Curriculum Organisation and Classroom Practice in Primary Schools* London: DES.
5   Kerry, T. (1982) *Effective Questioning* Basingstoke: Macmillan Education.
6   Crossman, M. (1984) 'Teachers' interactions with girls and boys in science lessons', reprinted in Kelly, A. (1987) *Science for Girls?* Milton Keynes: Open University Press.
7   McGlathery, G. (1978) 'Analyzing the questioning behaviors of science teachers' in Rowe, M. (ed.) *What Research Says to the Science Teacher Vol.* 1 Washington, D.C.: National Science Teachers Association.
8   SEAC (1991) *A Guide to Teacher Assessment: A Source Book of Teacher Assessment* London: Heinemann, Section 7.1.

FURTHER READING

Bentley, M. and Campbell, J. (1990) 'Assessing design and technology: Some possibilities and problems' in Bentley, M., Campbell, J., Lewis, A. and Sullivan, M. (eds) *Primary Design and Technology in Practice* Harlow: Longman.
Blyth, A. (1988) 'Appraising and assessing young children's understanding of industry' in Smith, D. (ed.) *Industry in the Primary School Curriculum: Principles and Practice* Lewes: Falmer Press.
Blyth, A. (1990) 'Some implications of assessment in primary industrial education' in Ross, A. (ed.) *Economic and Industrial Awareness in the Primary School* London: SCIP/Polytechnic of North London.
Conner, C. (1991) *Assessment in the Primary School* London: Falmer Press.
Hopkins, D. (1985) *A Teacher's Guide to Classroom Research* Milton Keynes: Open University Press.
Russell, T. and Harlen, W. (1990) *Assessing Science in the Primary Classroom: Practical Tasks* London: Paul Chapman.
Williams, P. (1990) *Teaching Craft, Design and Technology: Five to Thirteen* (Second Edition) London: Routledge.

Index